REMOTE SENSING
PRINCIPLES AND APPLICATIONS

2nd Revised Edition

A.N. PATEL
and
SURENDRA SINGH

SCIENTIFIC
PUBLISHERS

Published by:
Scientific Publisher
5 A, New Pali Road, P.O. Box 91
Jodhpur - 342 001 Rajasthan (India)
E-mail: info@scientificpub.com
Website: www.scientificpubonline.com

New Delhi
4806/24, Ansari Road,
Daryaganj,
New Delhi - 110 002 (India)

Dr. A.N. PATEL
INDORE- 452 008

Dr. SURENDRA SINGH
JODHPUR- 3420 011

3rd Print : 2024
Second Print : 2018
Second Revised Edition, 1999

ISBN: 978-81-72335-90-8

Printed in India

PREFACE

It is an established fact that for proper application of remote sensing techniques in the identification and mapping of natural resources, wastelands, temporal environmental changes, sand dunes, present and palaeo drainage systems and desertification, the knowledge of the basic principles of remote sensing is of paramount importance. Accordingly, an attempt has been made in this book to discuss the basic principles and applications of remote sensing.

In all, there are fourteen chapters, first eight chapters deal with basic principles of remote sensing and remaining three chapters highlights applications of remote sensing. In the chapters one and two (Introduction and Signatures in Remote Sensing), necessity, importance, scope and basis of remote sensing techniques have been highlighted.

The third chapter (Electromagnetic Radiation) deals with radiant energy from the sun, the electro-magnetic spectrum, atmospheric effects of radiation, absorpiton, transmission, reflection, atmospheric windows and black body radiation. The basis of signatures such as spectral, spatial, polarisation and temporal variations and methods for obtaining signatures based on laboratory, airborne and spaceborne sensors have been discussed in chapter four (Interaction of Electromagnetic Radiation with Matter).

Inchapterfive(SensorsUsedinRemoteSensing),spectral bands for sensors, classification of sensors and types of sensors needed for different spectral bands ranging from ultraviolet to visible, infrared, thermal infrared and microwave regions have been highlighted.

The types of platforms such as balloons, aircrafts, rockets and satellites for the sensors to record the images of terrain features have been dealt with in chapter six (Remote Sensing Platforms).

In chapter seven (Data Product), the type of remotely sensed data in the form of photographs, paper prints, satellite imagery, Computer Compatible Tapes (CCT ms) lmulti m concept in acquiring remote sensing data, advantages of Landsat imagery and conversion of data into information have been highlighted.

The chapter eight (Analysis and Interpretation Techniques)

deals with the image processing and enhancement of remotely sensed data for visual and digital analysis.

In chapter nine (Application of Remote Sensing in the Appraisal and Management of Natural Resources), the physical potentials and limitations of the natural resources such as landforms, soils, vegetation, landuse and water resources for regional planning have been discussed.

The chapter ten (Role of Remote Sensing in the Detection of Temporal Changes) highlights the temporal changes in saline areas, morphology of landforms, drainage systems, water bodies, grazing lands and forset cover. The types, extent and distribution of wastelands in India and Jodhpur district of Rajasthan and their management have been discussed in chapter eleven (Application of Remote Sensing in Wastelands Mapping).

In chapter twelve (Distribution and Management of Sand Dunes Using Remote Sensing Techniques), distribution and morphology of sand dunes, digital Landsat spectral characteristics of sand dunes, management and stabilisation of sand dunes have been discussed.

The chapter thirteen (Impact of Present and Palaeo Drainage Systems on Geoenvironment using Remote Sensing Techniques) highlights the positive and negative impact of present and palaeo drainage systems on Geoenvironment.

In chapter fourteen (Remote Sensing in Monitoring and Combating Desertification), the status of desertification, digital analysis of Landsat data for assessment of land vulnerability to desertification and measures for combating desertification have been highlighted.

It is hoped that this book will help the academicians, scientists, students, planners and policy makers in understanding the basic principles and applictions of remote sensing for rational regional development planning.

The authors wish to express their sincere thanks to Prof. Alam Singh for providing an opportunity to write this book. The sincere thanks are due to the Director and Head, Division of Resource Survey and Monitoring, Central Arid Zone Research Institute, Jodhpur for granting permission to the second author to write the book. The help rendered by the colleagues in writing, typing, tabulation, drawing and photography is thankfully acknowledged.

A.N. PATEL
SURENDRA SINGH

CONTENTS

1

INTRODUCTION

INTRODUCTION TO REMOTE SENSING

Every human being has a natural desire to know. The motivation to know what is behind the next hill, across the sea, through the forest, under the ocean, beyond the clouds or to observe the beautiful face of mother earth from space is a powerful goal to our restless quest for new horizons and fresh challenges. Man ms greatest technological breakthroughs have resulted from his curiosity about a phenomenon he could not experience directly with his senses. Therein lies the essence of remote sensing of environment.

Remote sensing means acquiring information about an object or phenomenon from a distance, i.e. without actually coming in contact with the object. Without direct contact, some means of transferring information through space must be utilized. A flow of energy, from the object being sensed to the sening device, is necessary. The quantity most frequently measured in present day remote sensing systems is the electromagnetic energy emanating from the object of interest (Fig. 1.1).

A great deal of energy is flowing around us. The most obvious is the reflected light, either natural or artificial. But a considerable amount of invisible additional electromagnetic energy is also present. To prove this, you need only turn on a radio or television receiver. These sensors will pick up the unseen waves emitted by a remote transmitters. These receivers transform radiation into a sound and/or image, which we can experience. We cannot perceive the waves because their wavelength and frequency are beyond the reception capabilities of human sensors. Human sensors have limitations. The human eye (remote sensor) can respond to light in the minute portion of electromagnetic spectrum i.e. between 0.4 and

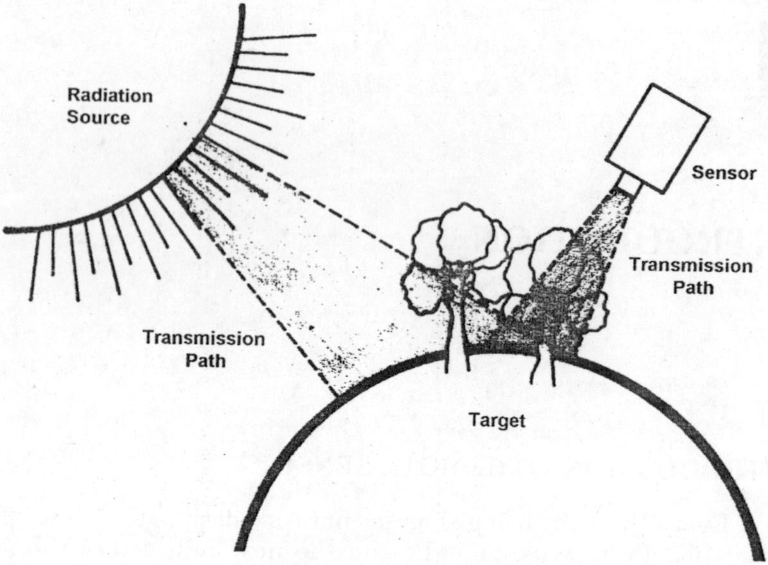

Fig. 1.1. Remote Sensing Model

0.7 micrometer wavelength. While the human ears can detect sound frequencies between 16 and 20,000 cycles/sec.

We live in a world full of radiation, some of it we can perceive, much of it we cannot. The job of remote sensing is to utilize this radiation to present an image which can be experienced by man ms more limited sensing capability.

NECESSITY AND IMPORTANCE

With growing population and rising standard of living, pressure on natural resources has been increasing day by day. It therefore, becomes necessary to manage the available resources effectively. For effective management of natural resources, renewable as well as non-renewable, it is essential to prepare accurate inventories of these resources periodically. Until recently such inventories were made almost entirely on the ground. Geologists travelled widely in exploring for minerals, foresters and agriculturist examined trees and crops in the forests and fields in order to assess their conditions, surveyors walked the countryside in the course of preparing the necessary maps. The advent of aerial photography represented a big step forward. In last few years, the making of aerial photographs has been augmented by a new technique called remote sensing, in which sensing is done simultaneously in several bands of the electromagnetic spectrum. In its fullest form, the technique ranges through the spectrum from the very short wavelengths at which gamma rays are emitted to the comparatively long wavelengths at which radar operates. In this way, much more information about an area can be secured compared to that obtained with conventional photography which is limited to the visible band of the spectrum.

APPLICATIONS AND SCOPE

Remotely sensed photographs and images are non-selective with respect to information content, i.e. the data interpreted from an individual record may be applicable to diverse fields. The information is useful to investigators in many diverse disciplines.

The oldest application of remote sensing technique is in military operation. Infra-red sensing is very useful in military activities as infra-red images can be passively made at night. Unlike radar, infra-red sensing does not require sending out energy to be reflected and perhaps detected by an enemy. Infra-red remote sensing is particularly useful in providing data that can be used to locate an enemy ms position, to estimate his numbers and to detect his movements. Movement or activity whether of bodies or of vehicles, generates extra heat, thus, an infra- red scanner carried by an air plane can gather a surprisingly large amount of information on enemy manoeuvres. Another major military use of infra-red remote sensing technique is missile guidance. Infra-red sensing is also very useful in detecting and mapping forest fires.

Geologists use remote sensing to find deposits of minerals and petroleum. Soil scientists can prepare inventory of the important characteristics of soil by intelligently relating them to geological and geomorphological features and the type of vegetation found on images obtained by remote sensing. Foresters and agriculturists can determine the kind of trees and plants which grow in an area, can assess their health and estimate their yield. Using remote sensing techniques, hydrologists can locate useful aquifers and can estimate the volume of surface and sub-surface flow in watershed. Oceanographers can map the movements of ocean currents, marine organisms and water pollutants. Geographers can analyse landuse patterns over large area and can study the effect of climate, topography, plant life, animal life and human activity in a particular area. Civil engineers planning large construction projects such as highways, airports, railways or dams can obtain data on landforms, rock materials, soils, type of vegetation and drainage condition in the project area. Remote sensing is invaluable to map makers in their efforts to identify ground features and to position them accurately.

Wild life managers can use aerial photographs for locating habitats of various kinds of animals. Violations of natural law often show up in photographs e.g. illegal mining in remote areas, pollution of water by illegal dumping of chemicals and effluents from industry, air pollution due to the release of the smoke and poisonous gases through chimanies of industries. The analysis of disasters like floods, fires and hurricanes can be assisted by the study of remote sensing data and the information so obtained can be used in making emergency decisions and in combating the situation. Fast development in industrialization throughout the world is causing the serious problem of environmental pollution. Monitoring of this can be effectively tackled by employing remote sensing techniques. Thus, remote sensing has multi-disciplinary applications. Now, remotely sensed data can be processed and interpreted automatically through fast digital computers and thus large volume of information can be dealt with rapidly.

Hence, for effective management of diminishing sources of natural resources and monitoring the quality of our environment, the remote sensing techniques have edge over conventional methods. It is certain that this emerging new technology has a very wide scope in future, as it provides timely and reliable information and if necessary repetitive also to the resource managers.

The principles and applications of remote sensing have been highlighted in different chapters of this book.

2

SIGNATURES IN REMOTE SENSING

BASIS OF REMOTE SENSING TECHNIQUES

The underlying principle on which the whole remote sensing technique is developed, is that *all objects on the surface of earth have characteristic spectral signatures*. A spectral signature of an object or a ground surface feature is a set of values for the reflectance or radiance of the feature, each value corresponding to the reflectance or radiance arranged over a different, well defined wavelength interval. In other words, it is distinctive set of distinguishable characteristics. The response of ground surface materials to incident radiation (i.e. reflectance) and\or the energy emitted by all objects as a function of their temperature and structure (i.e. emittance) essentially determine the signatures.

The knowledge of spectral signatures is essential for exploiting the potential of remote sensing techniques. This knowledge enables one to identify and classify the objects. It is also required for interpretation of all remotely sensed data whether the interpretation is carried out visually or using digital techniques. It also helps us in specifying requirements for any remote sensing mission e.g. which optimal wavelength bands to be used or which type of sensor will be best suited for a particular task. This is also necessary for analysing and designing sensor systems for specific applications.

Lillesand and Kiefer (1979) suggest to keep in mind that signatures, though quite often distinctive, are not absolute or unique in the real world. What we observe in the real world is the *spectral response pattern* rather than signature. The spectral response patterns observed by remote sensors may be quantitative but are not absolute. They may be distinctive but they are not necessarily unique.

Determinations of spectral signatures implies basic understanding of interaction of electromagnetic radiation with various earth surface objects.

3

ELECTROMAGNETIC RADIATION

INTRODUCTION

Radiation is one of the three commonly recognised modes of transference of energy, the others being conduction and convection. Radiation is unique among the three in that radiated energy can be transferred across free space as well as through a medium such as air. Thus, sunlight-a form of radiant energy crosses the gulf of emptiness between the sun and the earth. The radiant energy emitted is called electromagnetic energy in recognition of the fact that it has both electric and magnetic components.

Energy is propagated by electromagnetic radiation (EMR) with a velocity of 3×10^8 m/sec. EMR is one of the most useful force fields for remote sensing, forming a high speed communications link between the sensor and remotely located substances. Many remote sensing techniques utilize waves motion for the detection and identification of distant objects. Seismographs utilize seismic waves or stress waves in the crust of the earth, microphones and human hearing utilize sound waves in air, sonar sensors utilize sound waves in water and electromagnetic sensor utilize EM waves. EMR is a dynamic form of energy make manifest only by its interaction with matter.

Maxwell considered EMR on a macroscopic scale, the interaction with matter depending upon electric and magnetic properties of matter. His formulation imposed on limitations on the possible frequencies, wavelengths or amplitudes with which such radiation could occur.

The dynamic fields always occur together as inseparable partners, so that neither purely electric nor purely magnetic radiated waves will occur separately from the other. When EM waves are intercepted by matter, the result will depend upon both the magnetic and electric properties of the matter.

Radiant flux density of radiated waves is proportional to the squares of wave amplitudes. These waves propagate through empty space at a fixed velocity lC m. When waves propagate through material, the velocity of propagation depends upon the material properties and the frequency of the wave. In all cases, the relation between the velocity of propagation v, the wavelength l and the wave frequency f, is :

$$v = f.$$

As the frequency does not change when the waves penetrate matter, the wavelength must, therefore, change as the velocity of propagation changes. For instance, visible light propagates through glass with a velocity, lv m approximately C/1.5, so that wavelength of the wave in the glass is shorter than in free space by a factor of 1.5. This factor is called the index of refraction.

RADIANT ENERGY FROM THE SUN

Radiation, R_s, is emitted from the sun in accordance with the Stefan-Boltzmann Law :

$$R_s = E_s T^4$$

Where R_s = solar radiation

E_s = emissivity of radiating surface

= Stefan-Boltzmann ms constant 5.7 x 10^{-8} wm^{-2} K^{-4}

T = absolute temperature of radiating over surface, °K

The sun emits radiation as a black body (E_s = 1) at a temperature of 6000 °K in the spectral range of 1.2 to 10 micrometer. At shorter wavelengths, absorption by the sun ms atmosphere decreases the emissivity. The flux reaching the outer boundary of the earth ms atmosphere is approximately 2.0 cal. cm^{-2} min^{-1} and this value has been called the solar constant.

The wavelength of maximum emission intensity is given by the Wien Displacement Law,

$$_m = 2.898 \times 10^{-3}/T \text{ micrometer}$$

For the sun, the wavelength of maximum emission intensity is approximately 0.5 micrometer. Nearly 99% of the solar radiation falls within the spectral range of 0.15 to 4 micrometer. Approximately 30% of the solar radiation reaching the outer boundary of the earth ms atmospheric is reflected back into space by clouds and other atmosphere constituents. Approximately 17% of the incoming radiation is absorbed by the earth ms atmosphere and about 22% is scattered and reaches the earth ms surface as diffuse radiations. Therefore, only about 31% of the radiation entering the outer boundary of the earth ms atmosphere reaches the earth ms surface as direct solar radiation.

THE ELECTROMAGNETIC SPECTRUM

In 1800, the English astronomer Sir William Herschel discovered that if he moved a thermometer along the visible spectrum from violet to red, the temperature readings increased progressively. He also found that the temperature continued to rise into the dark region beyond the red, indicating the presence of radiation similar to light. A few years later, other investigators who were studying the chemical effects of light demonstrated the existence of radiation in the region beyond the violet end of the visible spectrum. Thus, it has been shown that the visible spectrum, the defined boundaries of which depend upon limitations of the human eye, is but a small, centrally located segment of a very much greater band of radiation, all of which travels with the spead of light and exhibits such light related characteristic as reflection, refraction, interference and polarization. This total range of radiation is called the electromagnetic (EM) spectrum.

There are three basic measurements which define the character of waves of the EM spectrum wavelength, distance from one wave crest to the next (l), wave velocity; speed at which the wave crests advance, wave frequency, number of wave crest which pass a given point in a specified period of time (f), cycles/sec. or Hz.

Because wave velocity is constant (the speed of light) throughout the entire electromagnetic spectrum, a reciprocal relationship exists between wave frequency and wavelength.

Fig.3.1 shows the extent of EMR spectrum. EMR, then presents a continuum of frequencies, wavelengths and quanta from long, very low frequency radio waves to the very short, high frequency gamma and cosmic waves. In practice, this continuum of electromagnetic energy, as seen in Fig. 3.1 is broken down into various wavelength bands or spectral regions such as x-rays, ultraviolet, visible, infra-red, etc.

The wavelengths which are much use in remote sensing are the optical wavelengths which extend from 0.3 to 15 micrometer. This is the region in which all objects on the surface of the earth, under natural conditions, either reflect incident solar radiation or emit

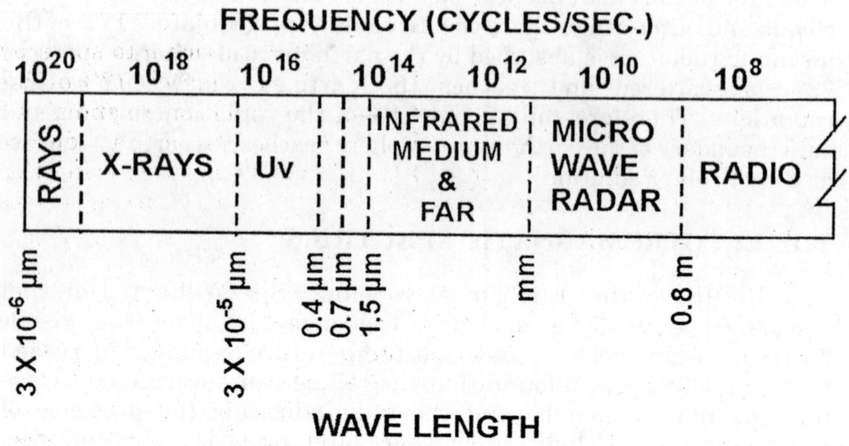

Fig. 3.1. Electromagnetic spectrum

radiations as a function of their temperature and molecular structure. The region between 0.38 and 3 micrometer is frequently referred to as the reflective portion of the spectrum. Energy sensed in these wavelengths is primarily radiation originating from the sun and reflected by objects on the earth. The reflective portion of the spectrum is further divided into the visible wavelengths and the reflective infra-red wavelengths. Since the human eye responds to radiation between wavelengths of approximately 0.38 and 0.72 micrometer, these wavelengths are referred to as the visible wavelengths. The region between 0.72 and 3 micrometer is referred to as the reflective infra-red portion of the spectrum, which in turn is further sub-divided into the near infra-red (0.72 to 1.3 micrometer) and the middle infra-red (1.3 to 3 micrometer) wavelength regions. Electromagnetic energy in wavelengths from 7 to 15 micrometer is in the far infra-red region of the spectrum. The portion from 3 to 7 micrometer has many atmospheric absorption bands and, therefore, the interpretation of the spectra becomes rather difficult. Windows falling within the 3 to 5 micrometer band have been used by some meteorological satellites. Such a band is difficult to use during day time because of the interference from radiations reflected by the earth ms surface.

Fig. 3.2 shows the optical spectrum. Dividing lines between different wavelength regions or bands, are arbitraily difined and should not be regarded as sharp, there being at each boundary a small overlapping range of wavelengths that might legitimately be placed in either group.

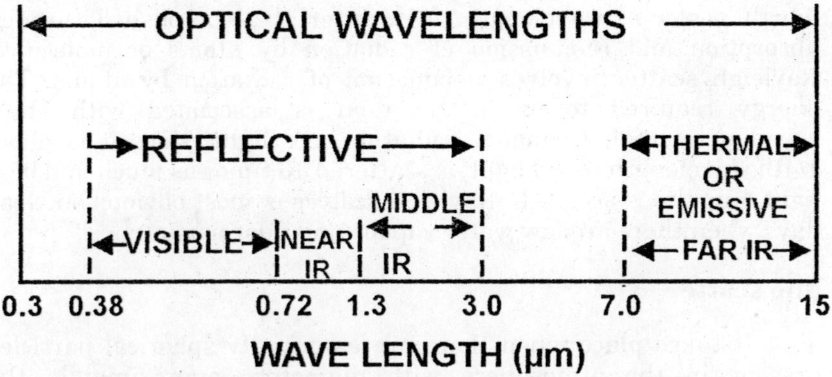

Fig. 3.2. Optical spectrum

The energy of a photon (quantum of electromagnetic radiation) or the radiant energy lQ m is given by the relation :

$$Q = h.f = h.c/$$

where, h = 6.625 x 10^{-27} erg-sec. (6.625 x 10^{-34} joule second) is the universal or Planck ms constant

This relation tells us that the energy of a photon varies directly with its frequency and inversely with its wavelength. The term photon is used to emphasize the quantized and statistical properties of radiation, while the term wave is used to emphasize the overall average effects of radiation.

ATMOSPHERIC EFFECTS ON RADIATION

Atmosphere may affect not only the speed of radiation but also its frequency, intensity, spectral distribution and direction.

Scatter

It differs from reflection in that the direction associated with scatter is unpredictable. The various type of scatters are Rayleigh scatter, Mie scatter, Raman scatter and non-selective scatter.

Rayleigh scatter

It is largely due to molecules and other very small particles, many times smaller than the wavelength of radiation under consideration. This type of scatter is inversely proportional to the fourth power of wavelength. All scatter is accomplished through absorption and re-emission of radiation by atoms or molecules. Rayleigh scatter involves re-emission of radiation by atoms. The energy required to excite an atom is associated with short wavelength, high frequency radiation. UV light, since it is about 1/4th wavelength of red light, is scattered 16 times as much and blue light four times as much. Rayleigh scatter is most obvious on clear days when there are few water vapour and dust particles.

Mie scatter

It takes place when there are essentially spherical particles present in the atmosphere with diameters approximately the wavelength of radiation being considered. The amount of scatter is greater than Rayleigh scatter and the wavelengths scattered are longer.

Raman scatter

It is much less common than the other types. It takes place when a photon has a partially elastic collision with a molecules, giving up energy to rotation or vibration of the molecule or gaining energy from the molecule.

Non-selective scatter

When there are particles in the atmosphere several times large the diameter of the radiation being transmitted, the scattering process is non-selective with respect to wavelength. Thus, water droplets, which make up clouds, scatter all wavelengths of visible light equally well, causing the cloud to appear white.

ABSORPTION, TRANSMISSION AND REFLECTION

Absorption

Certain wavelengths of radiation are affected far more by absorption than by scatter. This is particularly true of infra- red and wavelengths shorter than visible light. Absorption occurs when energy of the same frequency as the resonant frequency of an atom or a molecule is absorbed, producing an excited state. If, instead of re-emitting a photon of the same wavelength, the energy is transformed into heat motion and is subsequently re-emitted at a longer wavelength, absorption occurs.

The proportions of incident radiation which are involved in reflection, absorption or transmission depend partially on the radiant spectral distribution, the angle of incidence, the optical properties of the target (R.I. and absorption coefficient) and its thickness.

Transmission

Transmittance $T = \frac{I}{I_0}$

T = exp (j x) where exp = base of natural logarithum

= absorption coefficient

x = distance

Absorption and transmittance both vary with wavelength. This is basically a function of the natural resonance of the atoms, molecules, lattice, vibrations and free electrons which make up a substance. Thus, a target may behave very differently if exposed to radiation of different wavelengths.

Reflection

It refers to the process whereby radiation 1 lbounces off m m or object. Reflectivity of a surface is dependent upon :

(a) Angle of incidence

(b) The refractive index

(c) The extinction coefficient

The smaller the angle of incidence, the lower is the proportion of energy reflected. Reflectivity at any given angle is directly proportional to the refractive index. When light strikes a transparent medium at vertical angle, the amount of reflection is almost solely a function of the refractive index. Since, the extinction coefficient is proportional to the absorption coefficient, the greater the absorption the higher the reflectivity.

The greater the ability of a substance to absorb, the higher is its reflectivity and the lower is its transmission, all other things being equal. Absorption decreases at wavelength immediately above and below the maximum absorption bands of a substance for the shorter wavelengths, as absorption decreases reflectivity decreases and transmission increases, for larger wavelengths, both reflectivity and transmission increase over a narrow band width, then reflection begins to decrease.

ATMOSPHERIC WINDOWS

The solar radiation undergoes changes due to absorption, scattering etc. by the atmosphere before it reaches the earth ms surface. Almost all solar radiation of wavelengths shorter than 0.3 micrometer gets absorbed by the atmosphere it also absorbs some portion of the emitted terrestrial radiation. The most important constituent of the atmosphere, responsible for this are oxygen, ozone, carban-di-oxide, water vapour, etc. The region of the electromagnetic spectrum in which there is very less absorption of radiation are called the atmospheric windows. Fig. 3.3 shows atmospheric absorption of radiation.

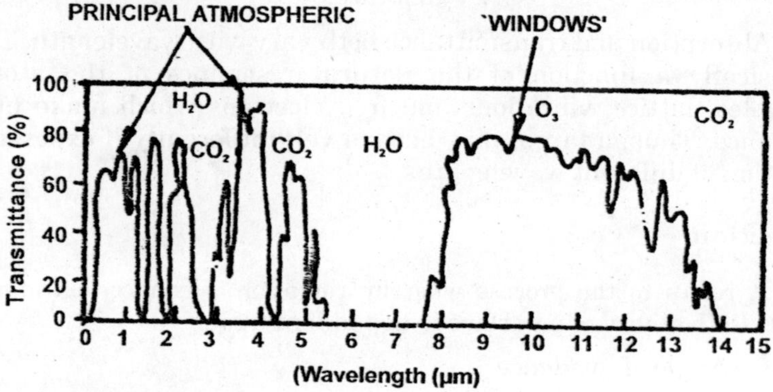

Fig. 3.3. Atmospheric transmittance of radiation

Table 3.1. Major atmospheric windows available for remote sensing

Spectral region	Spectral wavelength range (Micrometer)		
Ultraviolet & visible	0.30	j	0.75
Near infra-red	0.77	j	0.91
	1.00	j	1.12
	1.19	j	1.34
	1.55	j	1.75
	2.05	j	2.40
Middle infra-red	3.50	j	4.16
	4.50	j	5.00
Thermal infra-red	8.00	j	9.20
	10.20	j	12.40
	17.00	j	22.00
Microwave	2.06	j	2.22
	3.00	j	3.75
	7.50		11.50
	20.00	j	and above

It can be seen from the Table 3.1 that all the wavelengths shorter than 0.3 micrometer are closed for remote sensing while major principal windows lie in visible, infra-red and microwave regions.

BLACK BODY RADIATION

A useful concept widely used in radiation studies is that of a black body. A black body is defined as *an ideal object which absorbs all the radiation incident upon it and emits the maximum amount of radiation at all temperatures.* The total emitted radiation (radiant emittance) from a black body is proportional to the fourth power of its absolute temperature. This is known as Stefan Boltzmann Law and is expressed as :

$$W = KT4$$

Where W is emittance, K is the constant

$(K = 5.669 \times 10^{-8}\ W_m{}^{j2}\ 0_K^{j4})$.

T is temperature in degrees Kelvin

All bodies at all temperatures above absolute zero (-273°C or 0° K) radiate electromagnetic energy over broad range of wavelengths. Experimentally it is found that the wavelength near which most of the radiation is emitted depends on the temperature of the object. Fig. 3.4 shows black body radiation curves for various temperatures.

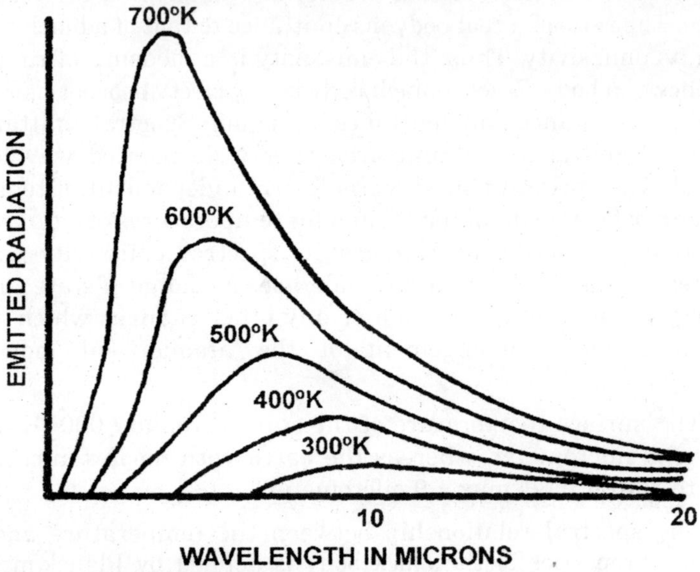

Fig. 3.4. Black body radiation curves for various temperature

It can be seen from this figure that as the temperature of a black body increases, the dominant wavelength shifts towards the short wavelength end of the spectrum.

Wien ms displacement law states that k

$$= \frac{C}{T}$$

where $_{max}$ is the dominant wavelength

C is a constant (C = 0.2898 cm °K)

T is temperature in degrees Kelvin

Temperature of a distant objects can be determined by measuring the spectrum of its radiation, which helps astronomers in detecting forest fire by thermal sensing and geophysicist in locating geothermal areas by infra-red sensing.

Since no real body is a perfect emitter, its emittance is less than that of a blackbody. Any body of material at a specific temperature emits radiation in accordance with its own characteristics. Spectral emissivity of the material is defined as :

$$E(\ ) = \frac{M \quad (\text{material},°K)}{M \quad (\text{black body}, °K)}$$

where M ( ) is spectral existance of a material

Existance is the flux density of radiant flux leaving a surface, while irradiance refers to flux density or radiant flux arriving at a surface. The ratio of a real body ms emittance to that of a black body is called its emissivity. Thus, the emissivity is a measure of radiating efficiency of a body. Since a black body is a perfect absorber as well as emitter, absorptance and emittance are equal. Spectral emittance is a radiant emittance per unit area over a designated wavelength interval. The spectral emissivity is nearly independent of temperature for most common materials and for temperatures of a terrestrial environment. Significant changes in spectral emissivity can be expected whenever the material undergoes a change of state such as melting, vaporization, oxidation or any other changes which alters the fundamental arrangement of the atomic and molecular components.

The surface temperature of the sun is around 6000 °K and l max = 0.5 micrometer, whereas the earth with mean temperature equal to 300 °K has l max = 9 micrometer.

The spectral relationship between the temperature and the radiation properties of the black body is defined by Planck ms Law which is as follows :

$$M = C_1 \; {}^{j5} \; [\exp(\frac{C_2}{T}) \; j \; 1]^{j1}$$

where M = Black body spectral existance ($W\, m^{-2}$)

C_1 = 3.74×10^{16} $W\, m^2$

C_2 = 1.44×10^{-2} m °K

= wavelength in meters

T = absolute temperature in degrees Kelvin

Many common materials are spectrally similar in 8 to 14 micrometer infra-red range, with a spectral emissivity between 0.85 and 0.95 micrometer. In the spectral range from 3 to 6 micrometer, the spectral emissivity of those materials tends to differ considerably. When a temperature measurement is made of a terrestrial material by remote means, it is wise to use 8 to 14 micrometer band, where it may be assumed that the emissivity of the various ground materials are similar and an estimated emissivity of 0.9 is probably not far from the truth. Chlorophyll in green plants absorbs visible radiation in the 0.4 to 0.68 micrometer range and converts most of that energy into chemical energy.

SPECULAR AND DIFFUSE SURFACES

Surface roughness is clearly a function of the wavelength of the incident electromagnetic radiation. Rayleigh ms criterion of surface roughness is given as k

$$h = \frac{\quad}{8 \ \mathrm{Cos}}$$

where h = height variations above a plane in wavelengths

= wavelength

= incidnet angle

What is important here is that wavelength of the incident EMR determines the roughness, e.g. in the radio range, rocky terrain appears smooth to the incident EMR, whereas, in the optical band, fine sand appears rough. In other words, a natural surface that appears very rough to the human eye may, in the radio range, appear specular and follow Snell ms law (where the angle of incident wave equals the angle of the reflected wave).

When the incident EMR wavelength is smaller than the surface height variations or the particle sizes that make up the surface (e.g. sand), the surface is diffused. EMR incident on such a surface is reflected in accord with Lambert ms Cosine Law of radiation.

A Lambertian surface is used, primarily in optics, to describe a diffuse surface that reflects the total EMR (visible) falling on the surface in all directions, the intensity of the reflected EMR is directly proportional to the intensity of the incident EMR times, the Cosine of the angle of incidence. The scattered EMR, or the radiance (W m^{-2} sr^{-1}) is usually defined in terms of an idealized diffuse surface emitter.

Another look at a diffuse surface is given by EM theory and Huygen ms principle. An infinitesimal surface dA is considered to be modelled by a large number of dipole oscillators. Radiation is emitted into the free space above the surface by these emitters. The pattern of radiation is a function of surface excitation. Typical radiation from a small area can yield a hemispherical or Cosine pattern. The EMR comes from the dipole oscillators where their contributions have random phase, therefore, no interference will occur and Huygen ms concept of each source sending out a spherical wave is applicable. Furthermore, a range of wavelengths is usually involved. Consider the small area dA emitting a multitude of wave fronts into the space above the area where their radii are much greater than the dimensions of dA. Upon inspection of a hemisphere of radiation, it is found that the intensity of EMR is greatest normal to dA, where Cos Q = 0 and progressively less for angles from the normal, because the radiation becomes less as the grazing angles are approached. It can be demonstrated that an area element D_a at a constant radius around dA will intercept radiation as Cosine q with a maximum normal to dA which in essence, is a statement of Lambert ms Cosine Law.

$$I(\ ) = I_0 \text{ Cos}$$

Where $I(\ )$ = Intensity as a function of angle from the perpendicular (to the surface)

I_0 = intensity at = 0

= angle from the perpendicular

It is clear that Lambert ms Cosine Law must involve many point source oscillators. The radiance for a Lambert surface, when observed remotely, is found to be independent of the angle , or in other words, it has a hemispherical distribution. This is the same as stating that the radiation from a diffuse emitter can be described in terms of an area seen by a remote sensor having a solid angle of acceptance (viewing angle) and positioned any place in the upper hemisphere. The radiance depends upon the projected area viewed.

Thus for a constant solid viewing angle, the surface area is the projected area divided by Cos . However, the intensity varies directly as Cos and the signal at any viewing angle is constant.

A surface is usually neither completely smooth or rough and the Rayleigh criterion provides a dividing line between truly smooth and rough surfaces. Consequently, in the analysis of reflection properties of various surfaces the sum of two components needs to be considered : a specular and diffuse component.

Specular reflection is that caused by a smooth surface and is highly directional. The glint angle of the sun, which can be deterimental to the viewing of aerial or space imagery, is good example of specular reflection. The specular component is phase coherent and the amplitude fluctuations are small.

Diffuse reflections or scattering is in contrast to specular reflections. It has little directivity for the EMR, is phase incoherent, and its amplitudes fluctuate in a random manner. Specular and diffuse reflections and its relation to surface roughness is shown in Fig. 3.5.

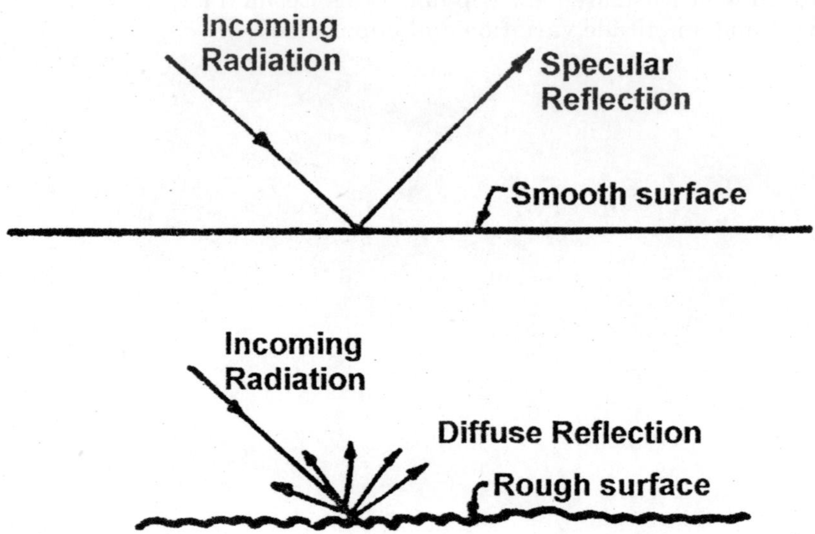

Fig. 3.5 Specular and diffuse reflection

Substantiation of the theory of reflection is evidenced by relating these two different reflections to the physically observed effects e.g. in the visible spectrum light waves are reflected from a mirror surface but scattered in all directions from a diffuse surface of matte white paper. For angles near vertical incidence, the matte white paper appears white showing diffuse scattering, however, at small grazing angles the phasing effects between the surface ray and below surface ray approach zero phase and specularity. Images are detected, however, the edges are dulled by the addition of a partially diffuse component.

The interaction of EMR with the earth or any substance which results in a reflection of a part of that radiation was shown to be quite complex when the variables of wavelength, angle, polarization, surface texture and the electrical properties of the material were considered. Nevertheless, certain basic characteristics of reflection have become useful tools for remote sensing experiments and data analysis.

Surfaces smooth by Rayleigh ms criterion reflect with high directivity, the angle of incidence equals the angle of reflection, with phase coherence, small amplitude change and with polarization. Surfaces rough by Rayleigh ms criterion reflect with a non-directive pattern which usually follows Lambert ms Cosine Law, with random phase and amplitude variation and unpolarized.

4

INTERACTION OF ELECTROMAGNETIC RADIATION WITH MATTER

INTRODUCTION

Interaction of electromagnetic radiation with surfaces causes the incident EMR to be reflected either specularly or diffusely, transmitted or diffracted into the surface, absorbed by the surface and sub-surface and all, or partially, re-radiated or scattered. All these processes are strongly dependent on the wavelength of the incident radiation as well as the atomic and molecular structure of the material and its condition.

$$\text{Total incident radiation} = \text{Reflection} + \text{Absorption} + \text{Transmission}$$

The amount of energy reflected back to a sensor depends on the wavelength and intensity of incident energy, the surface roughness, the orientation of the target and the target ms absorption characteristics with respect to wavelength. Reflection, transmission, absorption and emission are often treated as surface phenomena, however, a complete analysis must include penetration into the medium below the surface.

Everything in nature has its own unique distribution of reflected, emitted and absorbed radiation. These spectral characteristics can, if ingeniously exploited, be used to distinguish one thing from another or to obtain information about shape, size and other physical and chemical properties. In practice all remote sensing techniques are based on the observation of the reflectance of incident radiation and the emittance of radiation by the objects.

BASIS FOR SIGNATURE

The concept of discrimination is the basis for identification and classification of various earth resources from remote sensing methods. Discrimination is the simple recognition that different materials are indeed different. It also denotes the successive classification of larger classes into smaller ones, e.g. classification of land into agricultural and non-agricultural, the agricultural land into irrigated and non-irrigated, the division of irrigated area into area of different crops etc. Basis to such discrimination procedure is the concept of *signature*. There are four main characteristics of electromagnetic radiation which can be exploited for discrimination. They are :

(1) Spectral variations

(2) Spatial variations

(3) Polarization variations

(4) Temporal variations

All these variations are likely to be interdependent, e.g. shape (spatial) may be different at different time (temporal) or in different spectral bands (spectral). In addition, all the four types of variations are dependent on the angle of illumination and angle of viewing.

Spectral Variations

Spectral variations are the variations in the reflectance or emittance of objects as a function of wavelength. Colour of objects is manifectation of the spectral variation in reflectance in the visible region. Vegetation looks green because it reflects the *green* part of the incident light preferentially. Near infra-red reflectance of the vegatation is quite high. The functional behabiour of radiant intensity with wavelength can be different for different objects as shown in Fig. 4.1. The choice of spectral bands is intimately related to the spectral variations of various objects in such bands.

Spatial Variations

Spatial variations in the reflectance and emittance are attributes of the shape, size and texture of objects. The radiation emanating from different objects of a scene being characteristic of the optical properties of that object, provide the contrast between the object and its background. In the visible region of the spectrum, we are quite familier with spatial information provided by the shape, size, etc. as seen in photograph. However, in other regions of the

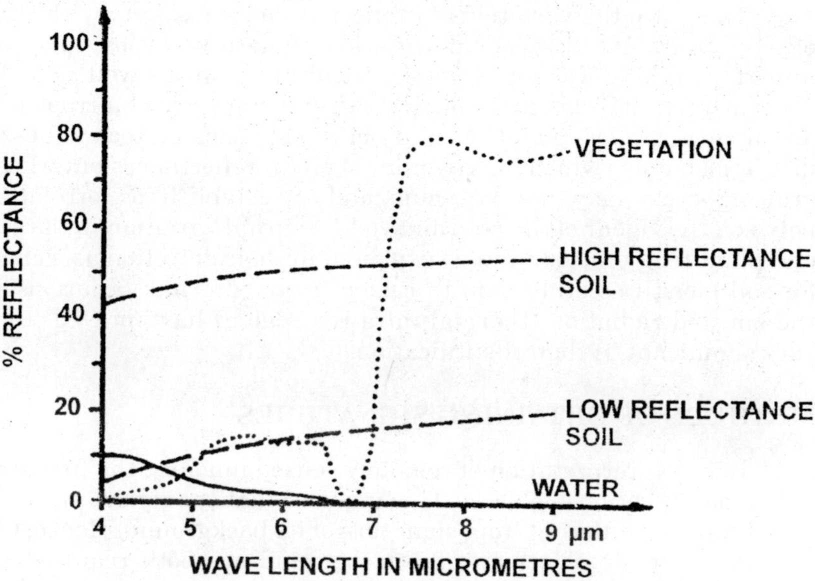

Fig. 4.1 Spectral reflectance curves for various objects

electromagnetic spectrum, the objects geometry may not furnish information about shape and size, e.g. a thermal scanner of hot spring will show the hot spots as per the temperature distribution across the surface of the hot spring. This may not show the same shape and size as a photograph in the visible region. The spatial reflectance of a crop, within a given wavelength band, will depend upon the size, and shape of the plants, their spatial frequency, field size and the scene background. A photograph taken from an aircraft altitude of large agricultural field will show an entirely different texture than shown by a photograph taken from orbital altitudes which may show nc texture.

Polarization Variations

Polarization is defined in terms of the orientation of the electric field component of electromagnetic radiation. Objects can be distinguished from each other by making use of the differences in the polarization of the reflected EMR. Such studies have become particularly important in the microwave region.

Temporal Variations

These are the variations in reflectivity or emissivity with time e.g. the diurnal and seasonal variations. These variations play an important role in the detection of agriculture crops as well as soils. The variation in reflectivity during the growing cycle of a crops has an important bearing on the ability of remote sensing techniques to distinguish crops which have similar spectral reflectances but whose growing cycle may not be same and to establish a correlation between the vigour of the crop and yield. Diurnal variation in the soil and plant temperatures can be exploited in their detection as well as for soil moisture estimation. In case of rocks, diurnal variations in the emitted radiation (thermal infra-red region) have proved to be very significant in their identification.

METHODS OF OBTAINING SIGNATURES

In any interpretation of remotely sensed imagery, the first step is the identification of objects. This requires that the signature of the object must be different from signature of its background. Hence, the signatures of the object and its background are both required for developing and employing discrimination techniques.

To make the remote sensing operational, the approach needed for establishing signatures of various surface objects of the Earth and their background is as follows. The approach involves :

(1) Laboratory studies

(2) Ground based field studies

(3) Airborne studies

(4) Spaceborne studies

(1) Laboratory Studies Under Controlled Conditions

The factor that can cause variations in signatures are so many that experiments undercontrolled conditions in the laboratory are essential. Such studies help in the basic understanding of the interaction of electromagnetic radiation with objects. Generally most of the laboratory studies on reflectance are made using spectrophotomters. This instrument can be used in transmission as well as reflectance mode. The reflectance studies make use of an integrating sphere with $BaSo_4$ coating on its inner surface. The laboratory instruments, in general, measure hemispherical reflectance (or emittance) and use rather small samples.

The main advantages of laboratory studies are : stable and controlled environment; no constraints of power, weight and space, speed at which observations are made is selectable. The disadvantage of such studies are : natural conditions under which remote sensing measurements are made may be quite different than laboratory environment e.g. study of vegetation specimen is usually carried out by removing them from the plants which disturbs the natural physiological conditions.

(2) Ground Based Field Studies

It is not possible to simulate all conditions prevalent in the field under the laboratory conditions. So, *in situ* measurements carried out under the natural conditions, play an important role. The *in situ* reflectance and emittance measurements can yield data which are more similar to the data gathered by airborne or spaceborne sensors. The advantages are : (1) measurements can be made under natural conditions of illumination and without disturbing the objects e.g. vegetation and soil, (ii) it is easier to make frequent spectral measurements under variable natural conditions and over a very long period of time, so that results can be effectively and economically applied to the multispectral data analysis. The disadvantages are : (1) portable and rugged equipment the *cherry picker* type platform is required, (2) unlike the laboratory, field measurements depend upon favourable weather conditions.

(3) Airborne Studies

Signatures obtained from the aircraft platforms are the most important for operational remote sensing. The platform being highly mobile, inaccessible areas can be covered. Signatures can be obtained using cameras with various film filter combinations, multispectral scanners, T.V. cameras, thermal scanners etc. These studies coupled with simultaneous ground truth data collection can furnish extremely reliable data for signature extraction. The disadvantages are : (1) more expensive for experimental studies, (2) measurements are not as precise as the laboratory or ground based field measurements, (3) small air-crafts are not *stable* and affects measurements, (4) sophisticated data reduction techniques are required as volume of data collected is very large. Data obtained using multispectral scanner is recorded on a magnetic tape in digital form. Signatures of various resources can be extracted using computer based processing techniques.

(4) Spaceborne Studies

The experimental spaceborne studies are conducted under most realistic conditions. Space provides an extremely reliable and quiet environment for the sensors. All the variables which can affect or modify signatures can be taken into account. However, there are constraints on the instrument regarding size, weight and power. The data have also to be telemetered to ground or data capsule has to be used, if the spacecraft is unmanned. Reliability of the sensors has to be extremely high as no maintenance is possible and failures are too costly. The calibration of the instrument has to be automated.

5

SENSORS USED IN REMOTE SENSING

INTRODUCTION

Remote sensors are mechanical devices which collect informations, usually in storable form, about objects or scenes while at some distance from them. All remote sensors simply record, in selected wavelength bands, variation in the amount of energy being reflected or emitted by various objects on the surface of the earth. An ordinary camera is probably the most familiar type of remote sensor. It utilizes visible light energy.

SPECTRAL BANDS FOR SENSORS

The electromagnetic spectrum is transparent at some interval called atmospheric windows and these windows are normally used for observing the earth. Sometimes the absorption bands are also used to study the absorption characteristic of the earth (Fig. 5.1). The spectral bands for visible and infra-red spectrum are :

0.4 - 1.1 micrometer (window)

2.0 - 2.5 micrometer (window)

3.0 - 5.0 micrometer (window)

5.7 - 7.2 micrometer (H_2O absorption band)

8.0 - 14.0 micrometer (window)

Similarly, the important microwave frequencies are given below. Band widths at these central frequencies are selected to get enough signal to noise ratio.

19 GH_z (window)

22 GH_z (H_2O absorption line)

35 GH_z (window)

The large spectral bands mentioned above are further subdivided into small discrete intervals and the observation is made in these small parts called spectral widths. The same area of the scene can be observed simultaneously in different spectral regions and more information about the characteristics of the objects can be obtained by this technique.

RESOLVING POWER

It is the ability of a particular sensor to render a sharply defined image. It is inherently poorer as we move to longer wavelength. This difficulty, to some extent is overcome by increasing the diameter of the optics, either lenses or antennas that are used to collect the energy. However, another factor that helps to compensate for loss of resolution is decreased atmospheric scattering at longer wavelengths.

CLASSIFICATION OF SENSORS

Sensors may be classified as active or passive, image or non-image forming, commercial or military (Fig. 5.1). An active sensor provides its own source of illumination, whereas the passive sensor does not. Photographic camera is a passive sensor when used in sunlight, while a camera used with flash bulb is active sensor. A radar and a laser altimeter are active sensors. Radar is composed of a transmitter and a receiver. The transmitter emits a wave, which strikes objects in the environmental field and is then reflected or echoed back to receiver. The chief advantage of passive systems are that they are relatively simple, both mechanically and electrically and they do not have high power requirements. Their disadvantages are that, particularly in wavebands where natural emittance or reflectance levels are low, high detector sensitivities and wide radiation collection apertures are necessary to obtain a reasonable signal level. Because of this, most passive sensors are relatively wide band systems. Another disadvantage is dependency on good weather conditions. Active systems require both transmitter and receiver units to produce imagery. High energy levels have to be generated, calling for more substantial power inputs than are necessary for passive systems. Their advantages are that they can obtain imagery in wavebands where natural signal levels are extremely low and they are independent of natural illumination for their operation.

Non-imaging sensors, in particular, are designed to give quantitative measures of the integrated intensity of electromagnetic radiation from all objects within their field of view as functions of time and wavelength. The imaging sensors, on the other hand, are designed to provide a visual image of their field of view either directly or indirectly through the stored information. Thus imaging sensors stress upon spatial resolution while non-imaging sensors stress upon time and wavelength resolution.

Classification of sensors as commercial or military is based on its application or purposes.

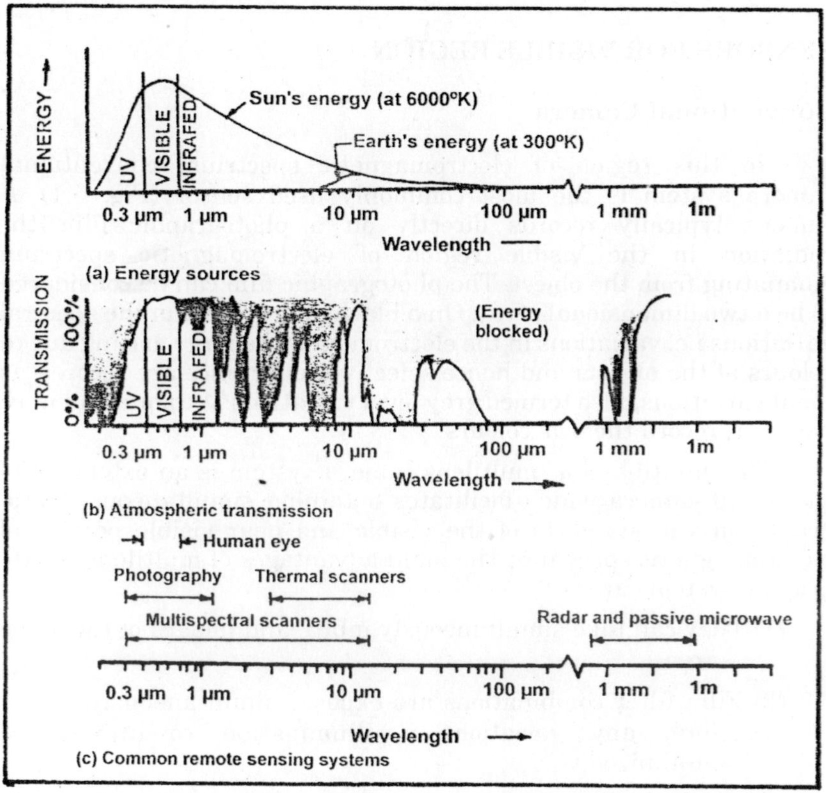

Fig. 5.1.Spectral characteristics of energy sources, atmospheric effects and sensing systems

SENSORS FOR ULTRAVIOLET REGION

Ultraviolet region is that part of electromagnetic spectrum which ranges from 0.4 to 0.01 micrometer. The UV spectrum may be divided into the near (0.4 - 0.3 micrometer), far (0.3 - 0.2 micrometer) and extreme (0.2 - 0.01 micrometer) (Fig. 5.1). The primary source of energy at these wavelengths is the Sun. Ultraviolet radiation from Sun is strongly attenuated by atmosphere at wavelength shorter than 0.28 micrometer owing to presence of ozone and molecular oxygen. Conventional cameras provide high degree of resolution and may be used in near ultraviolet region since film emulsions are available with sensitivities ranging down to 0.29 micrometer. But some lens materials are opaque for wavelengths shorter than 0.36 micrometer which limits the use of cameras unless with high quartz content are used.

SENSORS FOR VISIBLE REGION

Conventional Camera

In this region of electromagnetic spectrum, conventional camera system is the most commonly used sensor (Fig. 5.1). A camera typically records directly on a photographic film the radiation in the visible region of electromagnetic spectrum emanating from the object. The photographic film can be considered to be a two dimensional array. On a black and white film the spectral variations i.e. variations in the electromagnetic energy arising due to colours of the objects and hence selective reflectance are recored as tonal variations, also termed grey level variations. A colour film may, however, record the *true* colours.

The multiband or multilens camera system is an extension of the aerial camera which facilitates obtaining simultaneous photographs in various parts of the visible and near visible portion of electromagnetic spectrum. The main advantages of multilens photographic systems are :

(1) They can take simultaneously multiband images of the same area.

(2) Film filter combinations are exposed simultaneously, therefore, any variation in illumination conditions are minimized.

(3) Several film filter combinations can be exploited. The combinations which produce similar effect should be avoided.

Panoramic Camera

The panoramic camera is capable of taking high resolution photographs of a large area in a single exposure. This camera system provides unlike conventional camera, a high degree of image sharpness in every part of the photograph. This camera produces a narrow angular field and maintains a uniformly clear focus for the entire area covered as the film is exposed in the form of an arc instead of being kept flat.

Vidicon Camera

Vidicon or television camera works on the principle of scanning a photosensitive image plate by an electron beam. The spectral sensitivity of this camera ranges from 0.36 to 1.1 micrometer.

Major Components of Cameras

(i) Lens

(ii) Lens cone

(iii) Shutter and diaphragm

(iv) Camera body

(v) Drive mechanism

(vi) Film magazine

(vii) Focal plane

Aerial Camera Shutter and Characteristics

The aerial camera shutter is one of the main components of the aerial photographic camera and serves the function of preventing light from reaching the film emulsion except for brief interval during which the negative is exposed.

Shutter speed : It involves the duration of exposure, the importance of shutter speed and the ultimate goal.

Shutter speed accuracy : It is important in relation to photograph quality and use of colour film. The speeds of shutters are liable to change after considerable use and it is, therefore, desirable to test these speeds occasionally to obtain positive information according to their accuracy.

Shutter efficiency : It involves the efficiency of the between - the lens shutter and focal plane shutter.

Shutter life : Reliable life of the shutter is most important from a practical point of view. Life of a shutter is usually expressed in number of cycles of operation rather than as a guarantee over a specified period of time. In using the life cycle concept, it is assumed that the shutter is kept properly lubricated, cleaned, and free from corrosion.

Types of Shutters

Between - the lens shutter : This type of aerial camera shutter is the most popular and is widely used, particularly in the United States. The reasons for its popularity are found in its durability, mapping application, and absence of obstructions in the lens aperture.

Focal plane shutter : This shutter is so named because it operates close to the focal plane of the camera. This is also known as a curtain shutter because of the fact that it operates somewhat in the manner of a common window curtain. They are being used today primarily because they permit higher shutter speeds than can be had with the between - the lens shutter.

Louvre shutter : This type of shutter has not been used extensively in the United States, though in some countries its use is preferred to the between - the lens or focal plane type. The great advantage of these shutters lies in their application to reasonably high speed shutters for large lens apertures.

Shutterless strip camera : The major invention, perhaps in the field of aerial photography in the last half dozen years is the strip camera, and the corollary principle of image motion compensation (Katz, 1949). Remarkably low altitude photographs with image definition greatly exceeding any thing possible with conventional cameras and the highest speed shutters have been made in this camera. The greater advantage afforded by the strip camera is its ability to get good low altitude photographs under weather and light conditions which would necessitate shutter speed of about 1/50 second or slower.

SENSORS FOR INFRA-RED (IR) REGION

These generally image the emitted radiation, although reflected radiation may also be recorded. Since the thermal sensors do not depend wholly upon reflected energy for image formation, these systems provide both day and night capabilities. An optical mechanical scanner is the most commonly used imaging device that will sense the thermal infra-red (IR) portion of electromagnetic spectrum.

The infra-red portion of the electromagnetic spectrum is generally considered to be from 0.7 to 1000 micrometer (Fig. 5.1). In this range, IR radiation of a body is determined by two factors : (1) the nature of the objects surface - its emissivity and (2) its temperature. IR energy is recorded in two ways. The first is with radiometer, a device that records the radiations received and compares it to a fixed or known standard. The electronics of the system can then be used to generate an electrical signal based upon the differences between the standard reference and the object or scene being viewed. The other recording device is an infra-red scanner. The scanner is normally a system of mirrors which rotates or oscillates and focus the incoming radiation on a detector. The infra-red stimulation of the detector creates an electrical charge which can be amplified and recorded on a television tube, magnetic tape, video tape or photographic film. Using this kind of system, it is possible to get a picture of the thermal environment that we can experience with our normal human sensors. IR radiation can penetrate smoke or haze that is opaque to radiation at visual frequencies, however, it cannot penetrate rain or heavy fog. Qualitative thermal IR image is called thermogram which portrays in its grey tone variation, the general distribution of thermal radiation in the scene.

SENSORS FOR MICROWAVE REGION

All objects in the natural environment are emitting some microwave radiation. Even the atmosphere itself, through which this energy must travel to reach the antenna, emits radiation. Two properties of the atmosphere directly affect microwave detection. First, the atmosphere is a source of microwave energy and, therefore, illuminates all ground objects. Second, emission and transmission of the atmosphere place limits on the ability to detect ground objects. There are several atmospheric windows in the microwave region that are reasonably transparent and are *open* even during rather severe weather conditions. By using a sensor with a detector designed to receive wavelengths in one of these windows, it is possible to image ground objects remotely. The important properties of objects that determine the character of this radiation are emittance, transmittance, reflectance and object temperature. The character of external natural radiation sources that contributes to the reflected and transmitted radiation are also important. Selections of the wavelength for operation and the aperture size would be determined primarily by the resolution needed to map particular sources. Thus, if one assumes the use of the maximum aperture attainable at each wavelength, the wavelength to be used is determined by the resolution requirements, which dictate operation at the shortest possible wavelength.

Both active and passive sensing systems operate in the microwave portion of electromagnetic spectrum ranging from 1 mm to 1m (Fig. 5.1). They have the advantage of all weather capability since microwaves can penetrate through clouds and they can work day and night effectively. Disadvantages are large mass, power and volume they require and low spatial resolution.

Some imaging radar system are side looking i.e. they scan a path to one side of the aircraft ms flight line. Such systems are generally called *Side Looking Airborne Radar* (SLAR) system.

These systems work by directing a pulsed radar beam at right angle to the direction of flight of the aircraft. The returning signals reflected from the object on the ground are detected by the receiver aerial and converted to a light signal which is used to expose a film record analogous to the signal level. These systems offer day and night as well as all weather capability and ability to penetrate a cover of vegetation.

MULTISPECTRAL SCANNER (MSS)

An instrument which allows us to detect and record simultaneously both reflected and emitted radiation in several spectral bands, is the MSS. MSS operates on the principle of scanning successive lines at right angles to the flight path by means of a rotating or oscillating optical system. The radiation levels along the lines are recorded by appropriate sensor elements. When used in the visible band, the collected light can be split by the optics and separately filtered and recorded, giving simultaneous multispectral recordings from the one instrument. MSS can record in any part of the ultra-violet to near IR window. They are used also in the thermal IR windows (Fig. 5.1). Hard copy image records from line scanners are made by converting the original electronic signals into a fluctuation travelling light spot which is used to expose a film line by line. This is the simplest form of image data processing. It is quick, cheap and convenient. The number of density levels that can be recorded on the film are usually less than the number of signal levels steps that can be extracted from the record. This causes some loss of data. If the image data is initially recorded on magnetic tape, it can be replayed through the film processor at different settings to obtain a number of film records emphasizing different ranges of signal level. A further advantage is that by manipulating the data by using tape recording with ground replay and processing of the imagery, it is possible to obtain geometrically corrected and quantifiable calibrated imagery. Depending on the equipment, both digital and analog electrical- processing techniques are used. The line scan principle can also be used with microwave imaging systems, notably in the short wavelengths of less than 5 mm.

PHOTOMETER, RADIOMETER, SPECTROMETER, ETC.

Photometer

It is very sensitive dark room light meter with cadmium sulphide detector, with a spectral sensitivity sufficiently broad to cover the full photographic range.

Radiometer

A radiometer is a sensor that measures the intensity of electromagnetic radiation emanating from all objects within its field of view and wavelength range. The essential difference between photometer and radiometer, is that, former in general needs no internal reference of radiation, whereas radiometer do require such internal reference. In and beyond the infra-red portion of the spectrum, all objects, including the component parts of the radiometer itself, are sources of radiation, the contributions of these sources may be variable and hence result in error. Therefore, a stable controlled radiation reference must be provided.

Spectrometer

It is a radiometer to which has been added a dispersing element, so that the radiation from the target is measured as a function of wavelength or frequency.

Polarimeters

They are radiometers, photometers or spectrometers with the addition of a device to permit passage of single plane polarization only. Rotation of a polarizer around the optical axis of the radiometer controls the degree and angle of polarization of the received EMR.

RETURN BEAM VIDICON (RBV) CAMERA

The development of the vidicon camera has provided satellites with a method of obtaining high resolution pictures without the need for conventional photographic film.

The direct beam vidicon consists of an evacuated glass envelope containing an electron gun which faces a thin photo-conductive insulating layer (the target) which is deposited on the end window of the tube. An electron gun scans across the target surface and removes any postive charge, so stabilizing the target surface at zero voltage. An exposure is made in the conventional manner by means of the shutter and lens assembly. An image then forms on the target, the illuminated regions inducing a positive charge in proportion to be brightness of the reflected signal. This charged pattern is retained on the target surface until it is scanned a second time, in this case the magnitude of the pattern being measured and recorded.

The RBV differs slightly from the direct beam vidicon described above. It relies on the strength of the returning electron beam after it has neutralized the charge on the target as a means of measuring the charge pattern. The advantages are : (1) camera can operate in poorer illumination condition and (2) the resolution is generally higher than that associated with direct beam vidicon.

THEMATIC MAPPER (TM)

This sensor has better spatial resolution, spectral separation, geometric fidelity and radiometric accuracy compared to MSS.

The TM differes from MSS in five ways :

(1) The TM obtains data on both the west to east and east to west scans. This bidirectional approach reduces the mirror scan rate.

(2) The TM detector arrays are located within the primary focal plane of the instrument, allowing incoming light to be reflected directly on to the detectors without transmission through the fibre optics as employed in the MSS. This helped in minimizing loss in the intensity of the incoming radiation. But as the detector arrays for different spectral bands are spaced apart in the focal place by a small amount, the same point on the ground is not simultaneously scanned in all spectral bands. Therefore, precise band to band registration is an extremely important aspect of the subsequent data processing.

(3) The TM uses seven spectral bands in contrast to 4 for the MSS. The band designation, spectral range and principal application are given in Table 5.1.

Table 5.1. Band disgnation, spectral range and application

Band designation	Spectral range (Micrometers)	Application
1.	0.45 - 0.52	Water penetration
2.	0.52 - 0.60	Measurement of visible green reflectance
3.	0.63 - 0.69	Vegetation discrimination
4.	0.76 - 0.90	Delineation of water bodies
5.	1.55 - 1.75	Differentiation of snow and cloud
6.	10.40 - 12.50	Thermal mapping
7.	2.08 - 2.35	Geological mapping

(4) In addition to the increase in the number of spectral bands, the TM also has an increased radiometric sensitivity. The range of discreate radiometric levels which can be sensed is increased from 64 to 256.

(5) The spatial resolution of the TM is far better than that of MSS. In TM a pixel size of 30 m is used in all bands except band 7 which has a pixel size of 120 m. This contracts with the 79 m pixel size, associated with the MSS.

6

REMOTE SENSING PLATFORMS

INTRODUCTION

The base, stationary or moving, on which remote sensors are mounted is called a remote sensing platform. The remote sensing platforms range from a camera on a tripod or a *cherry picker* stand on the ground to balloons, helicopters, aircrafts, rockets and spacecrafts.

The choice of a particular platform is mainly, guided by the choice of sensors which in turn is guided by the objectives of the programme. However, other key factors in the choice of a platform are the payload, operating height, operating range, time and cost.

AIR BORNE PLATFORMS

Balloons

Balloons are rather inexpensive platforms and offer a great variety of shapes, sizes and performance characteristics. They provide large field of view and in general, require no power. They have low acceleration and hence delicate instruments can be mounted. They also exhibit low vibrations. Free balloons can provide a platform at intermediate altitude between those of aircraft and spacecraft. The greatest disadvantage of free balloons is that their flight trajectory may not be controllable and they too much depend on meteorological conditions, particularly the winds. This makes extremely difficult to predict whether the balloon will fly over the specific area of interest or not.

There are three main types of balloon systems :

1. Free balloons
2. Tethered balloons
3. Powered balloons

Free Balloons

A free balloon can lift thousands of kilograms of scientific payloads and can almost reach the top of the atmosphere. Unless a mobile launching system is developed, the flights can be carried out only from a fixed launching station e.g. in India at present, Tata Institute of Fundamental Research, Bombay has set up a National Balloon Facility at Hyderabad.

Tethered Balloons

These balloons have the capability of keeping the equipments at a fixed position for any desired length of time and thus useful for many remote sensing programmes.

Powered Balloons

These balloons require utilization of some means of propulsion to maintain or achieve station over a designated geographic location. Since these can be remotely controlled, guided along a path or hover above given area within certain limitations, they offer good potential as a platform for remote sensing applications.

Aircrafts

Aircrafts have been playing an indispensable role in the development of remote sensing techniques. No earth resources satellite programme can be undertaken without using a well equipped aircraft in the initial stages for testing the sensors and various systems and sub-systems involved in such programme. Practically every type of aircraft ever built has been used, at one of the other time as a remote sensing platform. The advantages of using aircrafts as remote sensing platform are :

1. High resolution of data recorded.
2. Possibility of carrying large payloads.
3. Capability of imaging large area economically.
4. Accessibility of remote areas.
5. Convenience of selecting different scales.
6. Possibilities of repetitive surveys.
7. Adequate control at all time.

However, due to limitations of operating altitudes and range, the aircraft finds its greatest applications in local or regional programmes rather than measurements on global scale.

ROCKETS

Rocket platforms are for one time observations only, the data is collected with continuous change in range, direction and illumination conditions. Except for one time qualitative or reconnaissance purposes, rocket platforms are not of much use in regular operational systems.

SATELLITES

Satellite platforms enable large area coverage at frequent intervals in uniform solar illumination conditions and are highly suited for applications based on synoptic measurements over large areas and where periodic observations over the same area are required.

A practical lower limit on satellite altitudes of approximately 300 km is set by atmospheric drag. The upper limit on altitude is set by booster thrust capability and the gravitational field of the earth. Two types of orbits which have been used in earth applications programme are :

1. Geosynchronous or stationary orbit.
2. Sunsynchronous orbit.

A Geosynchronous satellite is a satellite located in an equatorial posigrade (west to east orbit with an orbital period equal to the earth ms period of rotation). The altitude of such an orbit is approximately 35,900 km. The primary advantages of the orbit is that the satellite is stationary with respect to the earth. This allows continuous viewing of that portion of the earth within the line of sight of the satellite sensors. Disadvantages associated with this orbit include the extreme distance for earth viewing, the potential shortage of spacecraft parking space due to uniqueness or ordital requirements and the need to have several satellites to obtain global coverage.

The sunsynchronous orbit is an orbit selected so that the spacecraft passes over the same ground track at the same local time each day. This condition is satisfied by a family of near polar retrograde orbits with altitudes between 300 and 1000 km. Such orbits have the advantage that data taking is standardized with respect to sun angle. Since the sunsynchronons orbit is near polar, it provides nearly global coverage. The disadvantage of this type of orbit lies in the fact that diurnal effects at given points along the track are missed.

7

DATA PRODUCTS

INTRODUCTION

Aerial photographs are most commonly used as contact prints, either on paper or on transparent bases. The prints may also be ratioed to equate in so far as possible the scale of successive exposures, rectified to compensate for displacement due to tilt, or simply enlarged. A series of photographs may also be assembled into a single composite picture or mosaic. The data products thus are available in the form of aerial photographs, mosaics and imagers.

AERIAL PHOTOGRAPHS

Positive Transparencies

Prints made on film, glass or other transparent materials may be viewed with transmitted rather than reflected light and normally show finer detail and sharper definition than do paper prints. Positive transparencies are also known as diapositives. Positive transparencies should be used whenever the maximum amount of information is desired from the photographs. Colour photography, in particular, is usually viewed in the form of positive transparencies.

Photographic Emulsions

Black and white panchromatic : The emulsion consists of a very thin layer of light sensitive silver halide crystals. These crystals when exposed to light undergo a photochemical reaction which produces an invisible latent photographic image, further chemical processing or development releases this latent image. The spectral sensitivity of the emulsion covers the spectral range from 0.4 to 0.7 micrometers. The advantages are : good definition and contrast, wide exposure latitude, low cost, subtle textural variations can be identified and geometric patterns are more obvious. The limitation may be that sometimes difficult to interpret ground features due to the inability of the eye to distinguish between subtle differences in grey tones.

Black and white infra-red : The spectral sensitivity of this type of emulsion is from 0.4 to 0.9 micron which includes some part of reflected infra-red. Since infra-red radiation tends to be strongly absorbed by water, the boundaries between water and land are very clearly defined.

Maximum reflectance from vegetation occurs in the reflected infra-red region of the spectrum. Thus it may be possible to differentiate between different species of vegetation, e.g. between coniferous and deciduous trees. Black and white infra-red emulsions have improved haze penetration in comparison with conventional black and white emulsion. Sometimes detail is lost in areas of shadow due to excessive contrast.

Colour : The human eye can detect about 20 to 30 shades of grey on a black and white aerial photograph, while over 200 different colours can be discriminated on a colour aerial photograph. Thus a colour photograph provides much more information than the corresponding black and white photograph. Perception of colour depends largely on the relative amounts of the three primary colours, blue, green and red which are reflected by a particular object. Combining or adding the three primary colours, therefore, enables all possible colours, to be produced.

An alternative approach to produce a particular colour is that of subtractive colour mixing. In this case, rather than adding the three primary colours, the three subtractive primaries, yellow, magenta, and cyan are combined. Subtractive primary colours are so named because each is obtained by subtracting one of the primary colours from white light e.g., yellow is obtained by the subtraction of blue light, combination of the two remaining primaries red and green, produces yellow. Similarly, cyan is formed by subtracting red light and combining the remaining blue and green primaries. Colour aerial photography uses the second approach.

False colour infra-red : The spectral sensitivity of this emulsion ranges from 0.4 to 0.9 micrometer. Unlike *normal* colour emulsions, colour infra-red emulsions are designed in order to record green, red and infra-red energy. However, since the emulsion dyes which are used are the same as colour emulsions, i.e. yellow, magenta and cyan, the result is that the colours ascribed to their natural colours. The result, therefore, is a *false* colour image in which a blue image results from objects reflecting primarily green energy, green images result from objects reflecting primarily red energy and red images result from objects reflecting primarily in the near infra-red portion of the spectrum. The advantages are : (1) penetration to haze, (2) provides accurate identifiable data on vegetation, rocks, soils, water bodies and moisture distribution. But it has lower resolution than colour film and expensive than other films.

Types of Aerial Photographs

Aerial photograph may be classified according to orientation of camera axis, lens system and special properties.

Camera axis

(i) *Vertical* : Photographs whose optical axis makes an angle of less than 3° with the vertical or perpendicular (7.1a).

(ii) *Divided verticals* : Photographs taken simultaneously by the same aircraft with two cameras, of which one is vertical and the other is inclined at least 8° from the perpendicular.

(iii) *Low obliques* : All photographs, whose optical axis makes an angle greater than 8° with the perpendicular, but in which the horizon is not visible. (7.1b).

(iv) *High obliques* : Oblique photographs in which the horizon is visible (7.1c).

Lens system

(i) *Trimetrogon photograph* : The photograph taken with three cameras exposed simultaneously with two obliquely mounted usually at 60° to the vertical and the midle one vertically mounted (90°).

(ii) *Multilens photograph* : The photograph taken with a camera consits of two to nine lenses, all of which from either simultaneously or seperate negatve or one master negative.

Special properties

(i) *Continuous strip photograph* : Continuous strip photograph is taken with the aerial negative passing continuously over a narrow strip in the focal plane which admits the light.

(ii) *Infra-red photograph* : Infra-red photograph is a variation of the convectional black and white photograph. It is taken by using infra-red or minus blue filter with film which has been sensitised to the infra-red rays.

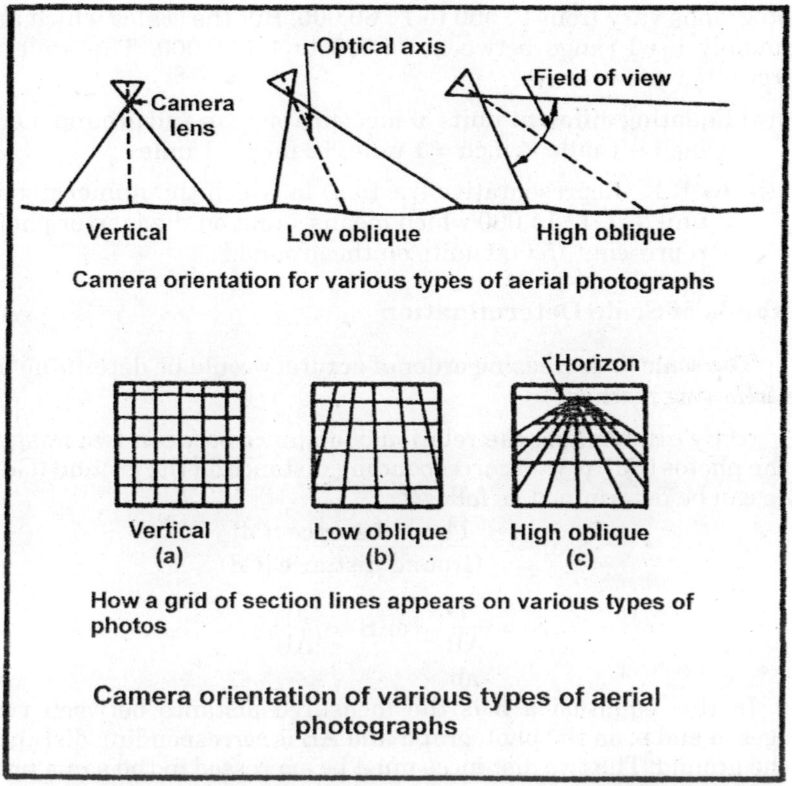

Fig. 7.1. Camera orientation of various types of aerial photographs

(iii) *Colour photograph* : The photograph taking by natural colour panchromatic film.

(iv) *Radar photograph* : Radar photograph using a radar screen is photographed on an ordinary film. Radar is active system of photography and the photograph can be taken at night or on cloudy or overcast days.

Size of Aerial Photographs

Aerial photographs are of different size viz. 12 x 12, 13 x 18, 18 x 18, 30 x 30, 24 x 48 centimetres. However, the aerial photographs of 13 x 18, 18 x 18 and 24 x 24 cm size are generally preferred.

Scale of Aerial Photographs

The scale is the relationship between distance on a map or photo and the actual ground distance. The scales of aer photographs vary from 1 : 500 to 1 : 60,000. But the scales which are commonly used range between 1:5,000 to 1 : 30,000. The scale is represented in two ways.

(a) Equating different units of measure on map and ground, i.e. 1 inch = 1 mile, 4 inch = 1 mile, 12 inch = 1 mile.

(b) As R.F. (Representative fraction) in which the numerator is unit, e.g. 1 : 15,000 which means 1 unit on the map or photo represents 15,000 units on the ground.

Methods of Scale Determination

The scale in decreasing order of accuracy could be determine by the following methods.

(i) By establishing the relation of distance between two images on the photos (Pd) as well corresponding distance on the ground (Gd), scale can be determined as follows :

$$\text{Scale} = \frac{\text{Photo distance (Pd)}}{\text{Ground distance (Gd)}}$$

$$S = \frac{1}{\frac{AB}{ab}} \quad \text{or } S = \frac{ab}{AB}$$

In this equation a b is the measured distance between two images, a and b, on the photograph and AB is corresponding distance on the ground. The two distances must be expressed in the same unit of measurement.

Example : Determine the scale of a photograph if the distance between two images on a vertical aerial photograph is 6 inches and the corresponding distance between the two objects on the ground is 2 miles.

$$S = \frac{1}{\frac{2 \times 5280 \times 12}{6}} = \frac{1}{\frac{126720}{6}} = \frac{1}{21,120}$$

Scale = 1 : 21, 120.

(ii) By establishing the relation of photo to ground with the help of a map :

If the distance between two images on a photograph which can also be located on the map, is measured, the horizontal measurements of these distances form a ratio, which when multiplied by the R.F. of the map gives the R.F. of the photograph. If lg m be the ground distance between two images, lm m the map distance and lp m the photograph distance then R.F. of map is m/g and R.F. of photograph is p/g.

$$\frac{\text{R.F. of photograph}}{\text{R.F. of map}} = \frac{p \;\; g}{m \;\; g} = \frac{p}{m}$$

$$\text{R.F. of photograph} = \frac{p}{m} \times \text{R.F. of map}$$

(iii) The scale lS m of a vertical aerial photograph is expressed by the following equation and also the figure 7.2.

$$S = \frac{1}{\frac{H \quad h}{f}} \quad \text{or } S = \frac{f}{H \quad h}$$

Where f is the focal length of the aerial camera, H is the flying height above sea level and h is the averge elevation of the terrain above sea level, or, if the scale of a specific image is considered, h is the elevation of the corresponding object. The values of f, H and h must be expressed in the same units of measurement. The result is a representative fraction corresponding to map scale.

Example : Determine the scale of a vertical aerial photograph if the focal length of the camera is 8 inches, the flying height is 14,000 feet and terrain is about 900 feet above the sea level.

$$S = \frac{1}{\frac{\frac{14000 \quad 900}{8}}{12}} = \frac{1}{\frac{13100 \times 12}{8}} = \frac{1}{\frac{157200}{8}} = \frac{1}{19{,}650}$$

Scale = 1 : 19,650.

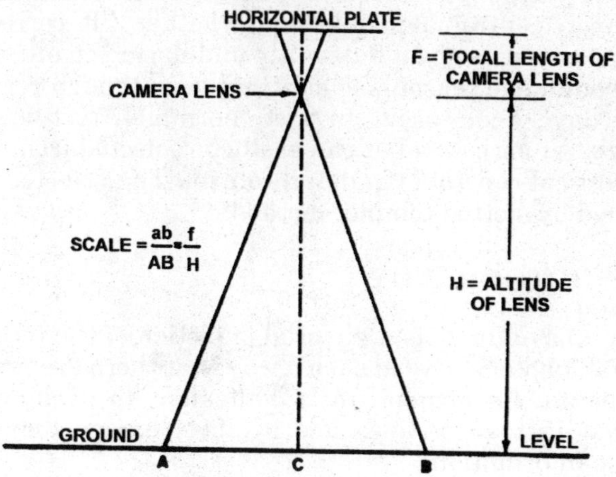

Fig. 7.2. Determination of scale

Types of Mosaics

(i) Uncontrolled mosaic : This is an assembly of photographs covering an area so that edges are matched as far as possible. This is the quickest and the cheapest method of obtaining topographical information about an area. The photographs can be assembled and then the mosaic can be photographically reduced to any appropriate approximate scale. For reconnaissance surveys even in mountainous terrain the mosaics could be justifiably used. An approximate scale can be indicated on the mosaic.

(ii) Controlled mosaic : This mosaic is prepared by providing enough planimetric control either by ground methods or by photogrammetric methods so that at least four control points one in each corner of that portion of the photograph which is to be used are available. The photographs are rectified i.e. the tilt displacement is removed. Normally this type of map should comply with standard map accuracy. This would be possible only when the terrain is flat or near flat as the relief displacement is to be limited to 0.25 mm on the scale of mosaic.

This mosaic is then as good as a map and would give as good results as a properly surveyed map. However, as is obvious the major limitation is that the terrain has to be near flat.

(iii) Semi-controlled mosaic : In between the two categories of mosaics mentioned above we have the semi-controlled mosaic which as the name siginifies, is a mosaic prepared with some control but not enough so as to achieve the accuracy of a controlled mosaic. *Normally because of terrain limitations most of the mosaics would fall in this category.*The accuracy of this type of mosaic would depend upon control available and whether tilt corrected photographs have been used. In flat/gently undulating country this type of mosaic, would give reasonable accuracy even though rectified prints may not have been used. In fact, normally, rectification is not restored to. As for control, it can easily be selected from the existing trigonometrical control/map detail control. The selected control can be extended by slotted templet method.

ORTHOPHOTOGRAPHY

It can be defined as a pictorial depiction of the terrain derived from aerial photography in such a way that there are no relief or tilt dispalcements. An orthophoto is equivalent to a planimetric map except that instead of lines and symbols, image tonal variations convey the information.

If the photographed terrain is flat and level, a rectified aerial photograph would be the same as an orthophoto. Compared to a palnimetric map, orthophoto offers several advantages. Since the details are reproduced in pictorial form, there is a wealth of information on the orthophoto, whereas in a line map the operator has already performed the classification task and selected only those features deemed important. Being an accurate map and rendering abundance of detail makes an orthophoto useful for more applications than either a planimetric map or a regular aerial photograph.

AERIAL PHOTOGRAMMETRY

Photogrammetry is defined as *the science or art of obtaining reliable measurments by means of photographs*.

Types of Aerial Photogrammetry

The aerial photogrammetry could be divided into three types (Whitmore, 1952).

(i) Ground or terrestrial photogrammetry

(ii) Aerial photogrammetry

(iii) Stereophotogrammetry

Ground or terrestrial photogrammetry denotes the use of photographs taken from the ground surface. Photographs taken on the ground with the optical axis of the camera horizontal are called horizontal photographs. Aerial photogrammetry connotes the use of photographs which have been taken from any airborne vehicle and these may be vertical or oblique photographs. Stereophotogrammetry means that overlapping pairs of photographs are observed and interpreted, or measured through the stereographic device which gives a three dimensional view and creates the illusion that the observer is viewing a relief model or terrain.

Major Problems of Aerial Photogrammetry

The difficulties in obtaining the precise measurements and compiling the accurate maps from the aerial photographs arise mainly from the following three sources (Whitmore, 1952).

1. Variation in conditions of photographing and developing

In order to obtain the perfect original photographs and to prepare the accurate maps in economical manner, the following ideal conditions are absolutely necessary.

(i) Each exposure should be made at the exact position and altitude selection in advance.

(ii) The camera should be exactly level with the optical axis of the lens exactly vertical at the instant of exposure.

(iii) The camera should be oriented in azimuth at the instant of exposure, precisely as planned in advance.

(iv) There should be no movement of the camera relative to the ground during the time of exposure.

(v) The camera lens should be free of distortion and otherwise optically perfect.

(vi) The camera should have stable and known metrical characteristics, and should be in perfect adjustment.

(vii) The negative film or plate should be perfectly flat and perpendicular to the optical axis of the lens at the instant of exposure.

(viii) The emulsion should be of uniform thickness, should give infinite resolution, and the film base should be stable in dimension.

(ix) Atmospheric conditions while taking the photographs should be ideal, free from dust, smoke etc.

2. Transferring information from the photographs to the map compilation sheet

The following operations involved in transferring and compilation of information from aerial photographs to the map, frequently cause difficulties.

(i) Developing the original films or plates.

(ii) Making positive prints from the negatives.

(iii) Operating the photogrammetric plotting instruments

3. Character of the earth ms surface

The character of the earth surface causes some difficulty in topographic mapping if the terrain is not flat. The photographic images of the aspects displaced due to effect of relief and tilt resulting in th distortion of the photographic images. However, the displacement of photographic images form the basis for determining the heights and contours from the photographs.

The distortion caused by the tilt and relief in the overlapping photographs could be rectified by photogrammetric methods using geodetic control points.

Elements of Photogrammetric Optics

In order to make proper use of photogrammetric materials in preparing topographic and other type of maps, the knowledge of optics is absolutely important. The salient features of the elements of photogrammetric optics are types of lens and lens aberrations.

The subject of the geometrical optics is based : (i) light travels in a straight line in a homogeneous medium; (ii) when a ray reflected from a surface, the angle of reflection equals the angle of incidence and both lie in the same plane; and (iii) when a ray is refracted (on passing from one medium to another medium of different density) the angles of incidence and refraction lie in the same plane and the size of the angle of incidence.

Types of lens

A lens may be defined as an optical medium bounded by spherical surfaces. The lenses are of two types :

(i) thin lenses; and (ii) thick lenses.

Lens abberations

An aberration is a defect in an optical image caused by the fact that the essentiality no lens system can form a perfect image. The different types of lens aberrations are briefly described below :

Spherical aberration: This type of aberration occurs when rays from various zones of a lens form at different places along the axis. It is caused by the spherical shape of the lens surfaces and decreases as the lens aperture is reduced.

Astigmatism: It is an aberration which causes a point object off the axis to be imaged as two mutually perpendicular short lines located at different distances from the lens. One of these lines is radial and the other tangential with respect to the centre of the field.

Longitudinal chromatic aberration: This type of aberration results when rays of various wavelengths or colours focus at different distances from the lens (the distance being measured parallel to optical axis).

Lateral chromatic aberration: This aberration is a difference in image magnification for various colours caused by chromatic aberration of oblique rays. Due to this defect, images near the outer portion of the field will be fringed with colour.

Lenses for photogrammetry

Different types of photographic lenses are designed, manufactured and tested for specific types of work. The aerial camera lenses used for photogrammetric work should have large relative apertures, large angular field, short focal lengths and high resolving power. The aerial camera lenses could be classified into three classes : (i) normal angle with fields upto 75°; (ii) wide angle with fields between 75° and 100°, and (iii) ultra wide angle with fields exceeding 100°.

Stereoscopic Parallax

Types of Parallax : If a nearby object is observed alternately with the left and right eye, its location will appear to shift from one position to another. This apparent displacement, caused by a change in the point of observation, is known as parallax. Parallax is a normal characteristic of overlapping aerial photographs, and it comprises the basis for three-dimensional viewing. The apparent elevation of an object is due to differences in its image displacement on adjacent prints. Two measures of parallax must be obtained in determining objects heights on stereoscopic pairs of photographs.

Absolute stereoscopic parallax (X - parallax) : It is the sum of the distances of corrosponding image from their respective nadirs. It is always measured *parallel* to the flight line.

Differential parallax : It is merely the difference in the absolute stereoscopic parallax at the top and base (ground level) of the object being measured. The basic formula for determining object heights or differences in elevation from parallax measurements is :

$$\text{Height of object } (h_o) = \frac{H \times dP}{P + dP}$$

where H = height of aircraft above ground datum

P = absolute stereoscopic parallax at base of object being measured

dP = differential parallax

If objects heights are to be determined in feet, the height of the photographing aircraft (H) must also be in feet. Once the exact photo to scale has been ascertained, the flight altitude can be found by multiplying the RF denominator by the camera focal length. Absolute stereoscopic parallax (P) and differetial parallax (dP) msut also be expressed *in the same units* ordinarily, these units will be hundredths of millimeters or thousandths of inches.

Differential parallax (dP) is usually measured stereoscopically with a parallax bar or stereometer employing the *floating mark* principle. Actually, the concept of differential parallax can best be illustrated by direct scale measurement of heavily-displaced images. If the height of a monument has to be calculated, the nominal scale of 1:4,800 is first corrected to an exact scale of 1: 4,600 at the base of the monument. A 12 inch camera focal length as used and the flying height above ground (H) is 4,600 feet.

Average photo base length (P) for the stereo-pair is 4.40 inches. Absolute stereoscopic parallax at the base and top of the monument is measured parallel to the line of flight with an engineer ms scale; the difference (2.06-1.46 inches = 0.60) is dP, the differential parallax of the displaced images.

Substituting the foregoing values in the parallax formula, we have :

$$h_o = \frac{4{,}600 \text{ x } 0.60}{4.40 + 0.60} = \frac{2{,}760}{5} = 552 \text{ feet}$$

Thus the height of a tree is 552 feet.

An accurate determination with an engineer ms scale would be virtually impossible and dP has to be measured under the stereoscope with a parallax bar or other *floating mark* device. Assuming a flying height of an aircraft above ground level 5000 m average photo base of 99.60 mm, and a differential parallax of 0.40 mm, the total height of a tree, for example, is calculated as below :

$$h_o = \frac{5{,}000\text{m x } 0.40 \text{ mm}}{99.60 \text{ mm} + 0.40 \text{ mm}} = \frac{2000}{100} = 20 \text{ m}$$

Thus the height of monument is 20 m.

DIGITAL IMAGE

In remote sensing, digital image is generally made from a collection of picture elements which are called pixels. This is like considering a photograph as large number of successive lines where each line contains a large number of square pixels. A number representing the average intensity of the scene within the square is stored. Thus, the image is a vast matrix of numbers. These images are very often held on magnetic tape and in particular on computer compatible tape (CCT) and the simplest way to store the data is to write all the numbers corresponding to one line of the image on to the tape as a *block* followed by a space and another block. This process is repeated until the last line in the image. A digital image can be read into a computer, line by line. It can be viewed on image display systems attached to computer. These images can also be *written* to photo or film writers. In this case, a light beam is caused to move in straight lines over a piece of unexposed film. The digital numbers along a line are used to modulate the light beam as it moves along line on the film. Then the film is developed. Thus the digital data is converted into photographic image (hard copy).

MULTI CONCEPT IN ACQUIRING REMOTE SENSING DATA

Multistation Photography

This term pertains primarily to successive over-lapping photographs taken along any given flight line as flown by a photographic aircraft or spacecraft. When two such photographs are studied in concert, the photo interpreter is better able to perceive features three dimensionally.

Multiband Photography

The specific composition of the energy emanating from a given type of feature (e.g. rock type, kind of soil, species of tree, or variety of crop) depends largely on the atomic, molecular and macromolecular composition of the feature. Consequently, each type of feature tends to exhibit a unique *total signature*. Thus, when brightness values seen in the series of photographs taken in different wavelength bands are suitably combined, it is possible to unambiguously identify specific terrestrial features.

Multidate Photography

A comparative analysis of photographs taken on a series of prescribed dates can provide additional elements of the identifying signature, because many types of features exhibit unique changes with the passage of time.

Multistage Photography

A multistage sampling scheme is one in which progressively more detailed information is obtained for successively samller sub-samples of the areas being studied. A three step process involving observations from space, air and ground is normally used (Fig. 7.3).

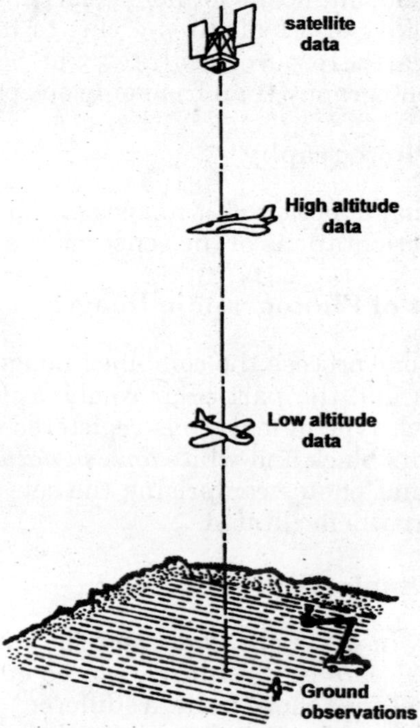

Fig. 7.3. Multistage remote sensing concept

Multipolarization Photography

When treated collectively, the various waves which comprise a beam of energy from the sun can be considered to be vibrating in all planes that are parallel to that beam as they travel through the atmosphere and impinge upon the earth ms surface. However, the light that is reflected back into the atmosphere from one kind of feature on the earth ms surfaces, such as a body of water, may be strongly polarized (i.e. vibrating primarily in one plane) while that reflected back from some other kind of feature, such as vegetation or fractured rock, may be polarized only slightly, if at all.

The simplest case of multipolarization photography is that obtained with two cameras, each equipped with a polarizing filter that preferentially transmits light waves that are vibrating in one particular plane. Cross-polarization photography is obtained when the polarizing filter of one of these cameras is oriented in one plane (ideally in a plane which includes the sun, camera and feature of interest on the ground) while the polarizing filter of the other camera is oriented perpendicular to that plane. In that event, strongly polarizing features appear much brighter as imaged with the first camera than the second, but not so for non- polarizing features. Since the polarizing capability of a feature is a clue to its identity, more information about earth resource features can be obtained from multipolarization photography than from only one polarization.

Multidirectional Photography

More information is obtained if images of the same area from one or more oblique orientations of the sensor also are available.

Multienhancement of Photographic Images

By additive colour process the combined images appear in full colour on the screen and the particular combination of hue, value and chroma with each type of feature is registered on the screen is directly relatable to its black and white *tone signature*, as portrayed on the three individual photos comprising the set. Photo- interpretation task is thus greatly facilitated.

Multidisciplinary Analysis

There are many instances in which more information about the total earth resources *complex* of an area can be obtained if two or more photo interpreters, each from a different discipline (e.g. forestry and geology) examine the photos, than if the interpretations were to be performed by representatives of only one discipline. This concept commonly is expanded, with the result that a multi-disciplinary photo-interpretation team used, is comprised for example of a geomorphologist, soil scientist, geologist, hydrologist, forester and land use analyst.

In order to exploit fully the advantages of this multi-disciplinary concept, the members of the team may find it profitable to use the *conference system* of photo-interpretation. They all study the same photograph at the same time, the representative from each discipline inturn contributes those portions of the overall analysis which he best can provide by virtue of his training and experience. The result is likely to be a much more complete and accurate analysis of the earth resources of an area than would otherwise have been possible.

ADVANTAGES AND LIMITATIONS OF LANDSAT IMAGERY

Main Advantages of Landsat Imagery

1. *Coverage of large area* : Each Landsat scene covers an area of about 34000 sq. km.

2. *Repetitive coverage* : Changes in land cover with time can be studied.
3. *Computer compatible data* : Landsat data can be digitally processed by computer in order to produce images that are easier to interpret. This improves the quality of interpretation.
4. *Availability* : Landsat data is easily obtainable from any of the regional remote sensing centres which exist in various parts of the world.
5. *Low cost* : The costs are relatively small when compared with the benefits which can be obtained from interpretation of the imagery.
6. *Spatial resolution* : This has been constantly improved. The latest SPOT satellite gives ground resolution of 20 m. It is capable of providing stereoscopic coverage of the earth ms surface, and hence height information is possible.

Main Limitations of Landsat Imagery

1. *Cloud cover* : In spite of the repetitive nature of Landsat, there are still several areas of the world for which there is only restricted coverage because of the problems of cloud cover.
2. *Lack of stereoscopic coverage* : It is not possible to get three dimensional view from Landsat products due to lack of stereoscopic coverage. Moreover, the delineation of the boundaries of the basins and sub-basins and interpretation and evaluation of their morphometric characteristics with reasonable accuracy is also not feasible.
3. *Misinterpretation of objects* : Similar spectral response from different objects and dissimilar spectral response from similar objects results into misinterpretation and wrong identification of features.

REFERENCE DATA OR GROUND TRUTH

Reference data are often referred to by the term ground truth. This term is not meant literally, since many forms of reference data are not collected on the ground e.g. ground truth may be collected in the air, in the form of aerial photographs used as reference data when analysing satellite imagery. Reference data may serve any or all of the following purposes :

1. To verify information extracted from remote sensing data.

2. To calibrate a sensor.
3. To aid in interpretation and analysis of remotely sensed data.

These data are often collected in accordance with the principles of statistical sampling.

CONVERSION OF DATA INTO INFORMATION

The *data* generated by remote sensing procedures become *information* only if and when some one understands their generation, knows to interpret them and knows how best to use them.

8

ANALYSIS AND INTERPRETATION TECHNIQUES

INTRODUCTION

Visual interpretation of pictorial image data has long been the work horse of remote sensing. Visual techniques make use of the excellent ability of the human mind to qualitatively evaluate spatial patterns in a Landsat scene.

Every one is a photo-interpreter. Almost every individual engages in image interpretation. Books, news papers, magazines, posters, movies and televisions all offer images to the observer, each image conveys ideas or impressions. These ideas or impressions are interpretations. Such interpretations may or may not be accurate and may be either conscious or un-conscious, partial or complete. Nevertheless, interpretation is an essential process through which information is obtained from images.

VISUAL INTERPRETATION

Image interpretation is defined as *the art of examining images for the purpose of identifying objects and judging their significance*. Interpreters study remotely sensed data and attempt through logical processes to detect, identify, measure and evaluate the significance of environment and cultural objects and patterns. Imagery of large areas permits the observer to perceive the relations of objects and their surroundings which to an observer on the ground might not be apparent. The image interpreter can examine objects, patterns and relationship in details and can compare these data with maps and other reference material. Stereoscopic viewing (creating the illusion of depth) enables the shape and height of objects to be determined. Perception of form makes it easier for the interpreter to identify objects and permits measurement of heights and slopes. Mirror stereoscope (Fig. 8.1) and Lens stereoscope (Fig. 8.2) are used for visual interpretation of aerial photographs.

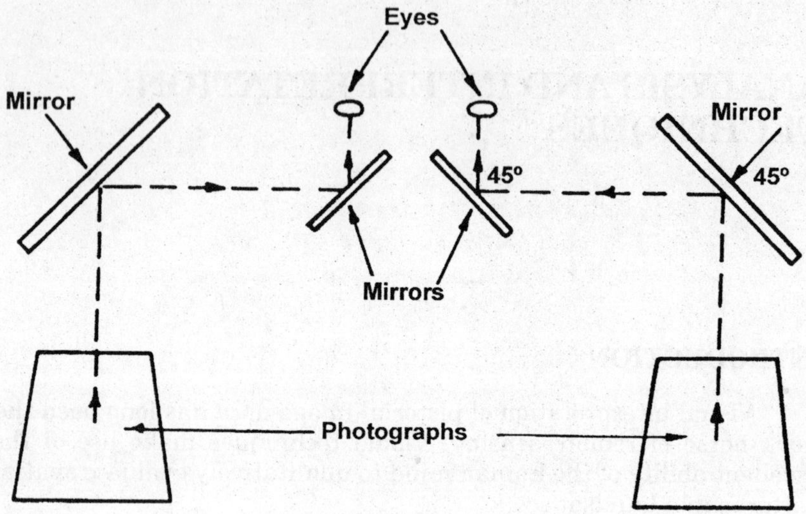

Fig. 8.1. Sketch showing the essential of the design of a mirror stereoscope

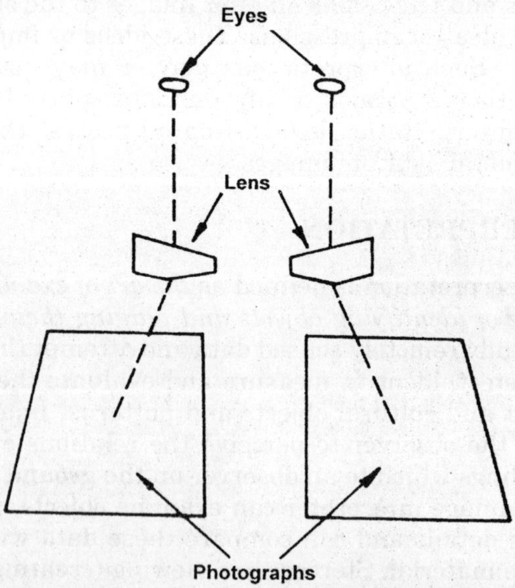

Fig. 8.2. Sketch showing the essential of the design of a lens stereoscope. The thin edges of the lenses are inside

The interpreter perceives reflected energy in a narrow band of the electromagnetic spectrum. Sensor systems and photographic emulsions currently in use, however, enable the interpreter to view picture like images of energy pattern beyond the limits of his vision. These energy relationships add significant information as the interpreter attempts to judge the significance of objects, phenomena, and relationships in environment. The permanence and fidelity of remote sensor imagery permit the interpreter to study intensively an area in a more leisurely fashion and in circumstances more favourable than may be obtained during direct observation.

Comparative studies of sequentially obtained imagery can also provide important historical documentation of trends in both physical and cultural processes. Image interpretation differs from direct observation in a real scope, perspective and temporal relationships.

Both image interpretation and statistics are techniques by which probabilistic statements can be made about object, phenomena, and relationships which may exist in our environment. The more knowledge the image interpreter has concerning the capabilities and limitations of the environmental and systems parameter which relate to the imagery, the greater the probability that his interpretations will be correct.

The image interpreter is often required to identify objects by the study of associated objects, or to identify object-complexes from their component objects. Most problems in image interpretation require the interpreter to have at his command knowledge derived not from the images themselves but from the relevant field or fields of study. The arrangements of particular systematic objects and phenomena recorded on remote sensor imagery facilitate interpretation because of the orderly way in which the objects have arrived on the earth ms surface. A geologic process, for example, gives rise to related features which possess the orderliness of common origin. Plant associations display the orderliness of common environment. Cultural features, associated by their common function, form industrial complexes (Fig. 8.3).

Basic Considerations in Visual Photo-Interpretation

Elementary characteristics

The following elements should be taken into consideration in photo-interpretation. Aerial photographs cannot be usefully employed without a knowledge of the characteristics marked on them.

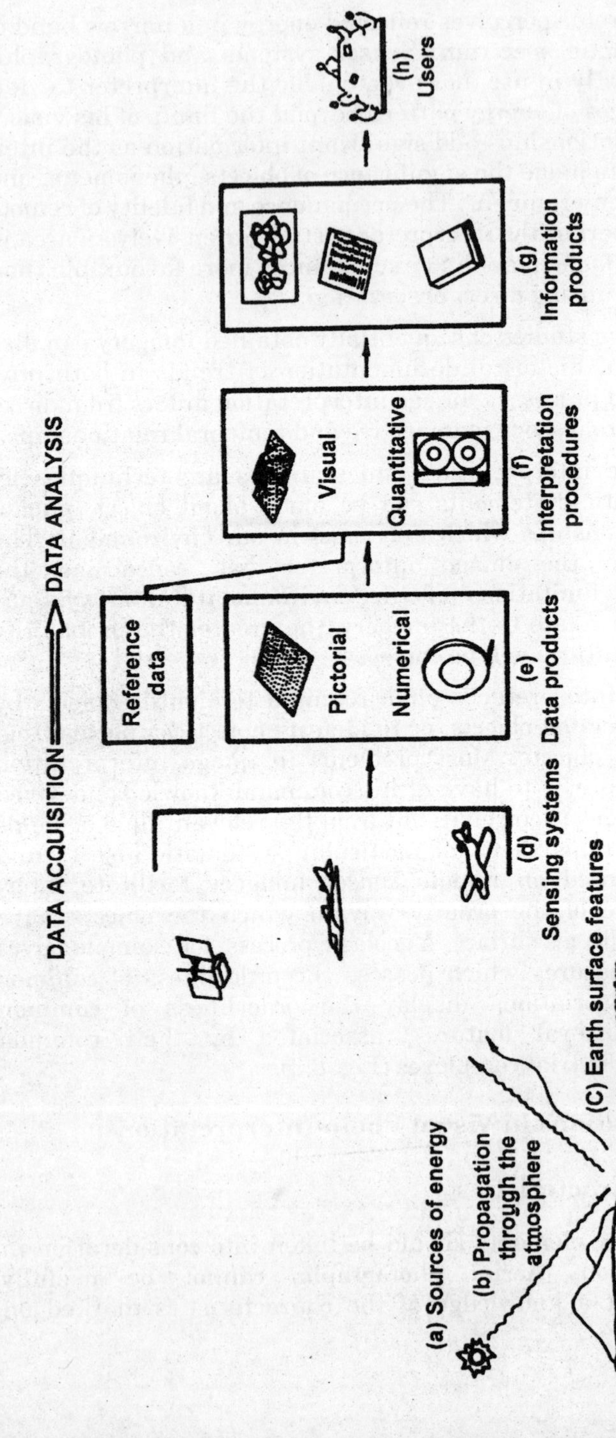

Fig. 8.3. Electromagnetic remote sensing of earth resources

These characteristics are :

(i) The nature of the mission : number, designation of the region (project area) surveyed, the organization which has taken the photos etc

(ii) The date of flight

(iii) The time of the day

(iv) The focal length of the camera

(v) The scale of negative

(vi) The aerial number of the photograph, indicating the run and the number

(vii) Area covered by the photograph and the photo index

(viii) Orientation of the photographs

(ix) Stereoscopic examination

(x) Calculation of scale.

(xi) Altitude, type of camera, film roll number, exposure number etc.

Besides the above characteristics, all aerial photographs carry four marginal and one central marks, which enable the intersection of vertical and horizontal lines, to locate the optical centre of the photograph, which is referred to as the principal point. The marginal marks are known as the fiducial marks. Recognition of these marks is of great importance for all further operations.

Elements of image interpretation

The basic elements of image interpretation are as follows :

(1) Size, (2) Shape and form, (3) Shadow, (4) Tone or colour, (5) Texture, (6) Pattern, (7) Site or location, and (8) Association and resolution (Fig. 8.3).

Size : The size of an object is one of the most useful clues to identify. By measuring an unknown object on an aerial photographs, the interpreter can eliminate from consideration whole group of possible identification.

Shape and form : The shape of some objects is so distinctive that their images may be identified solely from this criterion.

Fluvial, aeolian, glacial and marine landforms due to their distinct shape are conspicuously visible in three dimensions on aerial photographs and Landsat imagery. Stereoscopic photographs of glaciated mountains show erosional and depositional landforms like cirques, U-shape and hanging valleys, moraines and out wash plains; respectively with great clarity and beauty (Fig. 8.3).

Extrusive volcanic landforms are usually easily identified on aerial photographs if they have not been strongly modified by sub-aerial erosion. Cones and domes are the most spectacular forms produced by volcanic extrusion, they range in size from small cinder hills to huge composite cones consisting of interlayered ash and lava flow of innumerable eruptions.

Shadow : Shadows can help or hinder the interpreter, for they reveal silhouettes but hide some details. Shadows are familiar phenomena and in ordinary life we often judge the size and shape of objects or persons by observing the shadows they cast.

Shadows are particularly helpful if the objects are very small or lack tonal contrast with their surroundings. Under thses conditions, the sharp boundaries and shapes of the shadows enables the interpreter to identify objects that are just at the threshold of recognition.

Tone or colour : When the photo-interpreter understands the factors which govern photographic tone, he regards the tones or colours of objects of interest as major clues to their identification or composition.

Different objects reflect, emit and transmit different amount and wavelengths of energy. These differences are recorded as either tonal, or colour, and density variations on an image. Grey tone may be defined as any density or shade and including absolute black that is registered by the terrain on a exposed panchromatic aerial photographic negative or the print processed therefrom. True colour photographic imagery often facilitates interpretation by providing the interpreter with a more familiar view of the object under investigation. In black and white photography a water body may appear in tones ranging from white to black, depending upon the angle of the sun and the wave surfaces reflecting light to the camera lens. Soil tones, in general are more responsible to variations in moisture than to other natural factors.

Asthenumberofspectralbandsusedinremotesensingis increased, the identifying response pattern for each natural resource becomes more reliable and complete. Colour infra-red aerial photography and image enhancement devices, both optical and electronic, which can produce many different colour combinations from a set of black and white transparencies are being used more and more by image interpreters. The purpose is to enhance object to background contrast ratios and provide the interpreter with an expanded contrast between tone, with which he can more easily make his interpretation. In almost all cases it is the difference in tone or colour between objects or between an object and its background which is important. In fact, without a difference in tone or colour between the background and the edge of an object, there can be no detectable image. Various multispectral black and white images are reconstituted into a single colour composite image. The usual technique is to project each black and white image through a coloured filter. A battery of projectors is used so that all the images can be superimposed simultaneously on the screen. In such a composite image the tone or brightness of a ground feature as recorded in any given spectral band is used to govern the intensity of one of the colours used in the composite. By varying the selection of coloured filters, the analyst can change the colour contrast of the composite. Often by this means he finds that one combination of filters provides the best interpretability of one kind of feature, whereas other filters provide better interpretability of other features. The equipment used is called additive colour viewer.

Texture : Texture in images is created by tonal repetitions in groups of objects which are often too small to be discerned as individual objects. Texture, the visual impression of roughness or smoothness created by some objects, is a valuable clue in interpretation. On an aerial photograph, tree size is interpreted on the basis of apparent texture. The size of the object required to produce texture varies with the scale of imagery. In large scale aerial photographs, trees can be seen as individuals, their leaves or needles cannot be discerned separately, but contribute to the texture of the tree crowns.

Within a given range of scales, the texture of a group of objects (e.g. a timber stand of a certain species composition) may be distinctive enough to serve as a reliable clue to the identity of the object.

Pattern : Pattern or repetition is a characteristic of many man made objects and of some natural features. Outcrop patterns provide clues to geologic structure, and drainage patterns have orderly associations with structure, lithology and soil texture. Regional patterns, which formerly could be studied only through laborious ground observations, are clearly visible in aerial imagery. Drainage patterns like dendritic, trellis, radial, parallel, angular and rectangular developed on different types of landforms are clearly visible on aerial photographs and Landsat imagery (Fig. 8.4).

Cultural features : Some patterns in our environment are primarily cultural, others primarily natural cultural features are conspicuous in aerial images because they consist of straight lines or other regular configurations. Although, it is difficult to understand cultural features simply by looking at them, particularly from an altitude of several kilometers, the configuration seen from the air is often a sufficient clue to the function of the features.

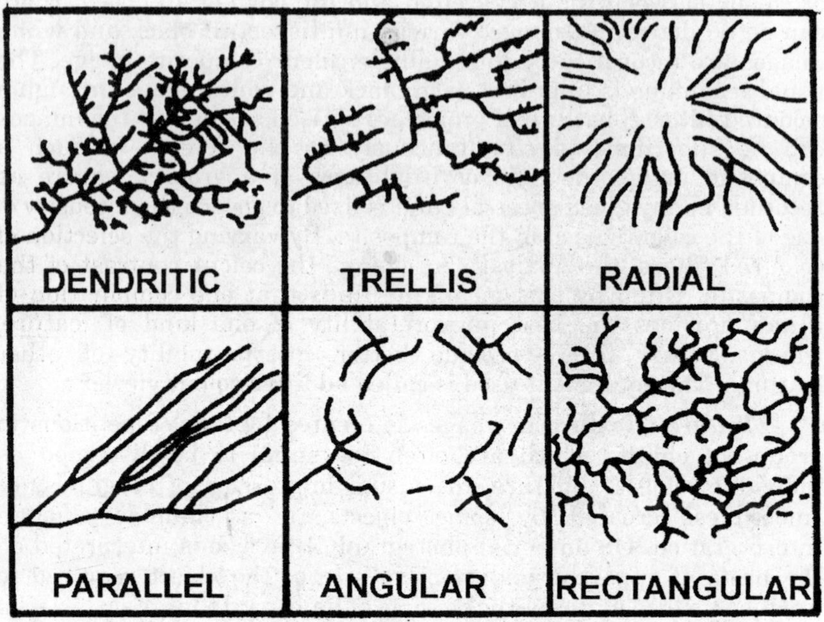

Fig. 8.4. Basic drainage patterns

Erosional features : Erosional features and evidences of accelerated erosion are important parts of the soil pattern. Gully cross sections and gradients are often useful indicator of soil texture. However, the climatic conditions under which the gullies were formed must be considered. In arid regions, intense runoff of short duration carves soft clay shales into bad lands. In humid areas where rainfall is distributed throughout the year, gullies in clay shales have softly rounded slopes.

Patterns of erosion are related to the work of wind, water, gravity and ice. Erosional features, particularly the forms of gullies are clues to soil texture and structure, profile development, degree of inundation and other conditions. Special erosional forms, such as the terracettes and pinnacles in loess and karst forms in limestone, reliably indicate the type and properties of the materials.

Gullies of various sizes, gradients and cross-sections develop in relation to soil texture, and the grain size and permeability of materials in an area can often be established by photographic inspection of the form, location and number of gullies. Short, close spaced, open gullies imply granular material. Both profile and cross section of a gully may show abrupt changes in slope.

Gullies in loess e.g. often show compound gradients, gullies in silty soils over clay pans have an abrupt change of slope in cross-section. The abruptness of the change in slope between gully sides and the surrounding ground is one of the best indicators of soil texture. Gullies in gravels have V-shapes, those in silts, U-shapes, those in clays softly rounded forms. The only gullies that may be expected to show vertical sides are those that are very young or those developed in materials with considerable strength because of cohesion and friction.

Site : The location of objects withrespect to terrain features or other objects is often helpful. Aspect, topography, geology, soil vegetation and the varied imprints of man ms culture are distinctive factors when examining a site. The relative importance of each of these factors will vary with local conditions but all are, to a greater or lesser degree important.

Association: Some objects are so commonly associated with other objects that one tends to indicate or confirm the other. Association is one of the most helpful clues to the identity of man-made installations.

Resolution : Resolution depends on many parameters, but it always places a practical limit on interpretation. The ability of an imaging system (including lens, filter, detector, emulsion, exposure, and processing as well as other factors) to record fine detail in a distinguishable manner is referred to as the resolution, or resolving power of the system. Resolution can be described in terms of ground resolved distance (GRD), line pairs per milimeter (LPM), modulation transfer function (MTF). If the image of the two objects are separable, the objects are said to be *spatially resolved*.

Study of patterns

A photo interpreter must tackle the following questions in the course of analysis and deduction :

(i) Is the pattern location related to : (a) land, (b) water, (c) regional, and (d) local?

(ii) Is the frequency of its occurrence : (a) continuing, (b) transient, (c) repetitious?

(iii) Is it seasonal, i.e. : (a) spring, (b) summer, (c) fall, (d) winter?

(iv) Is it : (a) common or (b) uncommon?

(v) Is it apparently : (a) natural or (b) cultural?

(vi) Is the pattern : (a) random or (b) systematic?

(vii) Is the tone of the pattern : (a) consistent, (b) lighter or (c) darker?

(viii) Does the unit vary in : (a) size, or (b) shape?

(ix) Is the pattern : (a) a unit in itself or (b) subject to breakdown of identity?

(x) Is the pattern : (a) near or (b) related to similar patterns?

(xi) Is it of : (a) the same, or (b) different datum?

(xii) Is it related to : (a) similar, or (b) dissimilar water bodies?

(xiii) Is it related to : (a) similar or (b) dissimilar landforms?

(xiv) To what major natural or cultural activity does it apparently relate :
(a) natural : (1) coastal waters, (2) lakes and rivers, inland waters, (3) forest and cover, (4) mineral, (b) cultural : (1) agricultural, (2) residential (3) commercial (4) industrial, (5) transportation, (6) storage, and (7) military.

(xv) To evaluate the effect of : (a) surrounding areas on tone, size, and shape of pattern, (b) distance or nearness on tone, size, and shape of pattern, (c) scale on image tone, size and shape of pattern, (d) tilt, bearing or dip on tone, size, and shape of pattern, (e) exposure, filters, or degradation on tone, size and shape of pattern, and (f) recording sequence on tone, size, and shape of pattern.

Factors of image interpretation

The basic factors of image interpretation are as follows :

(i) The quality of aerial photograph is an important aid to obtain reliable data. The property of the aerial photographs is apt to change due to technical factors like tone, film, paper, photographic chemicals, techniques of photo processing, printing and exposure. To eliminate such unavoidable defects it is necessary to reduce or intensify the tone of photographs for normal contracts for accurate phot-interpretation.

(ii) The scale of the aerial photographs is of prime importance in photo-interpretation. The photographs of 1 : 15,000 and 1:20,000 scales are well suited for vegetation, land use, forestry and geological photo-interpretation and that of 1:31,000 and 1 : 40,000 scales have been accepted for the interpretation of landform, soil and also for soil-landscape relationship.

(iii) The primary knowledge of the terrain factors of the area like soil, rock, vegetation, soil moisture, geological events, drainage patterns, ground relief, depositional and erosional landform features is essential to obtain reliable data from serial photographs.

(iv) The climatological and meteorological factors like atmospheric haze, clouds, sun angle and the seasons of the year are also to be taken into consideration for photo-interpretation.

(v) The effects of the tilt and tip on the shape of the ground object, effect of relief on the displacement of the image of the ground object, variation in the flying height causing scale error and also the chromatic and spherical lens aberration are sometime avoided for systematic image interpretation.

Methodology and Various Stages in Image Interpretation

The art of image interpretation depends on deductive and inductive methods. The image interpretation is the critical examination of the photographs followed by field check, depending on the type of survey and the degree of details required. The image interpretation involves the following stages :

(i) Indexing and rapid scanning of the photographs and to lay them into respective runs to get a compositve view of the area to be surveyed.

(ii) Selection of right and left hand photographs to avoid pseudoscopic effect.

(iii) Placing the photographs on a photo carriage at a proper distance.

(iv) Determination and transfer of principal points on each stereo-pair and to join them to get an air base.

(v) Setting up the mirror stereoscope with the eye base of the instrument parallel to the stereoscopic base of the photographs to get a perfect stereo-model.

(vi) Preliminary image-interpretation of the photographs for identification and deduction of significance of the images of ground objects.

(vii) Selection of query points and marking of traverse routes in different photo patterns.

(viii) Field traversing along the selected query points to collect the data on various geographical aspects and to compare the data already collected from the photographs.

(ix) Final image interpretation, preparation of photo legend and analysis of samples collected from the field.

(x) Transfer of data from the photographs into the base map on the required scale with the help of Zeiss Sketchmaster and Radial-line method.

(xi) Preparation of final maps using appropriate symbols and writing of report.

Convergence of Evidence

The completeness and accuracy of image interpretation are proportional to the interpreter ms understanding of how and why images show shape, size, tone, shadow, pattern and texture, while an understanding of site, association, and resolution strengthens the interpreter ms ability to integrate the different features making up a terrain.

There are many clues to identify unknown object. None of the clues is infallible by itself, but if all or most of the clues point to the same conclusion, the conclusion is probably correct photo-interpretation, then is actually an art of probabilities. Few things are perfectly certain in photo-interpretation in the way that one and one certainly make two, but many interpretations are so probable, when all visible evidences have been considered, that they may be safely regarded as correct. The difficult part of photo-interpretation consists in judgement of degrees of probability.

Visual interpretation techniques require extensive training and are labour intensive. Spectral characteristics are not always fully evaluated in visual interpretation efforts, because of the limited ability of the eye to discern tonal values on an image and difficulty for an interpreter to simultaneously analyze numerous spectral images.

IMAGE PROCESSING

Need for Image Processing

On film a limited number of grey levels can be held. Due to noise in the photographic material, it is only possible to have between 20 and 30 discrete and separable levels. Landsat MSS data has 64 grey levels and the Landsat Thematic Mapper has 256. The human eye can separate only 10 to 15 levels or about twice that if the grey levels are infact in large patches touching each other in sequence, as in a *grey wedge*. Thus, a film is incapable of holding all the information, and the human is unable to take in all the information to his *image processor* in his brain. Moreover, film is limited in its spectral response to the visible part of the spectrum and just into the near infra-red. Scanners can observe much more of the spectrum i.e. near infra-red, many regions in the mid infra-red, the thermal infra-red and the microwave region. There are maximum three layers on a photographic film and hence there can only be three independent images displayed together. Similarly there are only three primary colours and hence it is only possible to convey three images to the eye to be corrected. Landsat MSS has 4 spectral bands, the Thematic Mapper has 7 and many aircraft scanners have even more than 10. An image processing system is, therefore, required to correlate this data (Fig. 8.3).

Systems

The simplest type of system for image processing is a Computer. Images are read into the system, the required operation is performed using the Software and the results are written out on to another tape, or to a film writer, or displayed on a Television screen. Image processing deals with restoration, enhancement, geometric correction, matching, feature extraction, image models, texture analysis.

Restoration can be defined as reconstruction of an image towards the object (original) by inversion of the degradation it has undergone.

Enhancement deals with manipulation of data for improving its quality for visual interpretation.

The orbiting satellites send enormous data in the form of pictures of earth surface that are being used every day for resources monitoring. The acquired data can be used for various applications in the field of geology, geomorphology, hydrology, agriculture, forestry, land cover mapping etc. This data needs to be improved in its quality. They are to be corrected for geometric distortions, stored and retrived. Thus, the field of image processing deals with manipulation of data which is inherently two dimensional in nature. The field of image processing is developed recently i.e. in last 10-15 years. This is due to the development of large computers and advancement in image input and output devices like film recorders, scanners, flying spot scanners etc (Fig. 8.3).

Pre-Processing of the Data

No imaging system will reproduce images with accurate geometry or shape of the object. The reproduce image are always distorted. These distortions may be due to the characteristics of the sensing device itself or may be due to some factors like rotation of the earth, curvature of the earth etc. Therefore, before the data is recorded on computer tapes or on precision films, some corrections are required. These corrections may be radiometric or geometric. Radiometric corrections for MSS or TM are required for compensation of variation in detector responses within each band and for radiometric errors. These corrections are carried out based on statistical samples collected and calibration wedge data recorded during acquisition.

Image Averaging

Image averaging is a typical operation used in image processing to remove random noise (snow) from an image. Image averaging sums the brightness values of two or more frames on a pixel basis and then divides by the number of frames summed. The orignal consists of the desired video signal plus the added random noise. These random noise values may either increase or decrease the digital value of the desired video signal and appear in different spatial positions from frame to frame. When the system sums the signal values, the random noise values tend to cancel but the desired video signal increases. The final division scales the sum back to a range that can be displayed on a televison monitor. The actual increase in the signal-to-noise ratio is proportional to the square root of the number of images averaged if the noise is completely random. The result of averaging two or more frames of the same noisy image is a much clearer picture.

Contrast enhancement : In visual interpretation, in order to make maximum use of the available information, some sort of enhancement is usually needed. Image enhancement emphasises in viewing of the image for extraction of the information that may not have been so readily available in the original. Many enhancement techniques are available in practice. Principal among them are contrast stretching, filtering techniques, pseudo-colour composites, band ratioing, etc.

Contrast stretching : Adjustment of contrast has been done on photographic image since the origin of photography, but the problem of contrast stretching is much more extreme in digital images. The range of intensity values in an image is determined by a histogram operation i.e. by counting the number of pixels of each intensity. In histogram, all 256 possible intensity levels are represented on the horizontal axis while the vertical axis shows the relative number of pixels from the scene at each intensity. Fig. 8.5 shows a typical histogram.

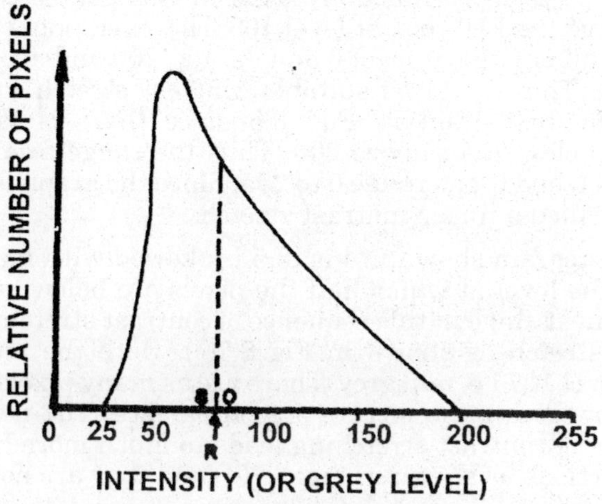

Fig. 8.5. Typical histogram of grey level in image

The histogram shows that the intensity values or grey levels range from 25 to 200. This range is stretched to the range of 0 to 255. In essence, this makes the lightest pixel white, the darkest pixels black and linearly arranges the intermediate intensities to span the range of 0 to 255. This transformation can be done by lookup tables (LUT) on the frame grabber. To understand the action of a LUT, consider that the LUT is a small block of memory that receives on 8-bit address an input and outputs the 8-bit data stored at that address. The input addresses are pixel values and the values stored at these addresses are the transformed values that are output. Thus, rather than doing the transform on each and every pixel, the system checks the LUT to find the pre-calculated value. This saves time since for example every pixel with an intensity value of 200 will have a transformed value of 140. Any arbitrary intensity transform can be generated and loaded into the LUT. Fig. 8.6 shows block diagram of the operation of LUT.

Fig. 8.6. Block diagram of the operation of a look up table

Fig. 8.7(a) shows what happens to the data initially. Data entering the LUT at level 100 leave the other side at level 100 as well. This graph is called *transfer function*. Since it shows how the input data are transferred to the output. In Fig 8.7(a) there is the *unity transfer* function or the one that does not change the data. In Fig. 8.7(b), the transfer function has been changed so that data at level 100 leave the LUT box at level 109. However, note that data at 25 or below, all emerge at level 0 and data at 200 and above, emerge at level 255. This would be suitable contrast stretch for the data shown in the histogram in Fig. 8.5 since there is virtually no information below 25 or above 200. Thus the range from 25 to 200 has been stretched to the range 0 to 255. Since the graph is a straight line, this is called a linear contrast stretch.

The histogram shown in Fig. 8.5 is skewed with median at 80 (Median is the level at which half the pixels are below and half are above). The next stage in this sequence of contrast stretching is a two step linear stretch as shown in Fig 8.7(c). Here the median data emerge at level 127 i.e. mid grey. There are as many pixels above this level as below giving the picture a good balance. The next stage in the sequence of contrast stretching leads to much more complicated transfer functions as shown in Fig. 8.7(d) and they are non-linear.

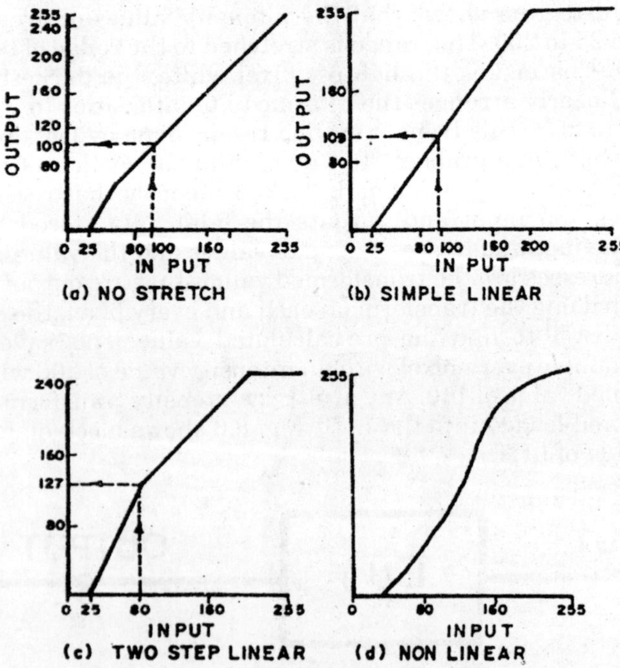

Fig. 8.7. Transfer functions

It may be easier for the eye to see some features in the more contrasty image, but others will be lost in the highlights and shadows. This demonstrates that a single contrast stretch for all applications is a compromise and that the human eye can extract most information for a particular application by using an interactive image processing system to produce a specific stretch.

Edge enhancement : Another class of image processing operations uses the information from the area surrounding each pixel to change the spatial contrast of the image. When an area operation is used to sharpen or enhance the edges in an image, it is called an edge enhancement. Edge enhancement can use convolution to implement a spatial filter that accentuates high frequencies.

The convolution operation consists of multiplying pixel in the *neighbourhood* of a given pixel by constants, summing and scaling, and replacing the center pixel by the result. The constants used for the multiplication are called the Kernel of the convolution.

Density slicing : Enhancement can also be carried out by assigning various colours to various intensity ranges in the imagery. In other words, image features corresponding to different spatial frequency ranges can be assigned various colours. This process is known as density slicing. If required pseudo-colour composites can also be produced using single band imagery.

Classification : The user marks an area on an image displayed on a screen, with a box or by running a line round it. He should have ground truth for this area i.e. he should have information from a field-trip to this part of the area in the image. The user knows what is actually there on the ground. This known area is then used by the processor to classify the whole displayed image, and the user assesses the accuracy of the results by comparison with the areas for which he has ground data. Further training areas can be added until the classification best fits the ground data, where it exists. Thus, the classification process allows the user to extrapolate from known small test areas to a whole region, with some degree of confidence, and hence to produce maps, statistics and so on.

Band ratioing : It is often used in practice for enhancement of multispectral data. Dividing one band of an image by another tends to show the spectral characteristics of land units irrespective of illumination conditions e.g. grass on the sun-lit side of a hill would appear much brigther than grass on the dark side in the original image, but both sides of the hill would appear very similar in an image formed by the ratio of the two bands. If one image was subtracted from another of the same area, in the same spectral band, but taken at different time, the regions of change will be highlighted e.g. an urban development, a change in vegetation forest area etc.

9

APPLICATION OF REMOTE SENSING IN THE APPRAISAL AND MANAGEMENT OF NATURAL RESOURCES

INTRODUCTION

It is an established fact that planning for integrated development of any region whether within administrative framework (i.e. a district) or a natural framework (i.e. a basin) depends upon the reliable and comprehensive appraisal and integration of natural resources. The appraisal of natural resources such as landforms, soils, vegetation, land uses and water resources through conventional methods is unreliable, time consuming and costly. Remote sensing techniques through visual interpretation of aerial photographs, satellite imagery and digital analysis of Computer Compatible Tapes (CCTs) in conjunction with ground truth have been proved faster and economic for the appraisal and integration of natural resources. Although Landsat images in the form of black and white and colour transparencies and computer compatible tapes have been found very useful for this purpose, the IRS satellite images due to their higher resolution ranging from 5.8 to 72.5 m seems to have much better potentialities.

Keeping in view this practical application of remote sensing techniques, the studies on the mapping of the distribution of natural resources, their appraisal and management by using visual and digital analysis techniques in conjunction with ground truth have been done by several workers in different parts of the country and abroad. The results of such studies have been reviewed under this chapter.

Landforms act as an environment for the development and distribution of soils, growth of vegetation and circulation of surface and ground water. Sufficient work has been done on the genesis and morphological characteristics of the landforms in different parts of the country by using conventional methods. But the conventional methods have been proved costly and time consuming and the accuracy of classification and mapping of landforms is also not satisfactory. Recently developed aerial and space remote sensing techniques have been found economical, faster and accurate for the classification and mapping of landforms. Landforms being most conspicuous features are distinctly visible on the aerial photographs and satellite imagery. The environmental hazards like wind erosion/deposition, water erosion, salinity/alkalinity and waterlogging associated with different landforms could also be easily identified and mapped for systematic and rational environmental planning. The present and prior drainage channels and drainage patterns can be identified and mapped from remotely sensed data.

The studies on different aspects of the landforms by employing remote sensing techniques have been done in various morpho-climatic zones. The salient features of these studies, emphasizing the use of remote sensing techniques in evaluating the geomorphological characteristics of the landforms have been discussed.

Classification and Mapping of Landforms

The use of aerial photographs of 1 : 25,000 and 1 : 30,000 scale has been made by Ghose and Singh (1965), Singh and Kaith (1971) for the classification and mapping of the landforms of Sumerpur and Pali blocks of Pali district. In these blocks, erosional and depositional landforms viz., hills and rocky/gravelly pediments, upper piedmont plains, flat older alluvial plains and younger alluvial plains were identified and mapped. Such landforms distinctly appear in three dimensions under stereoscope in white to dark grey and dark tone on aerial photographs. Water erosion and salinity/alkalinity hazards associated with various types of landforms could also be identified and mapped.

Based on Landsat black and white and false colour composite images and selected field checks, Jodhpur district located in western Rajasthan has been divided into ten major landforms (Singh, 1983). Due to sharp break of slope and distinct variations in tone varying from white to grey and dark, the hills, rocky/gravelly pediments and buried pediments could be easily separated and mapped from each other on LANDSAT band 5 image (Fig. 9.1), Flat and sandy undulating older alluvial plains and younger alluvial plains and graded river beds distinctly appear in white to grey and dark grey to dark tone on Landsat band 5 image (Fig. 9.2). Sand dunes and interdune plains due to their forms, patterns and location could be separated and mapped from Landsat band 7 image. The parabolic, longitudinal, transeverse and obstacle dunes distinctly appear on

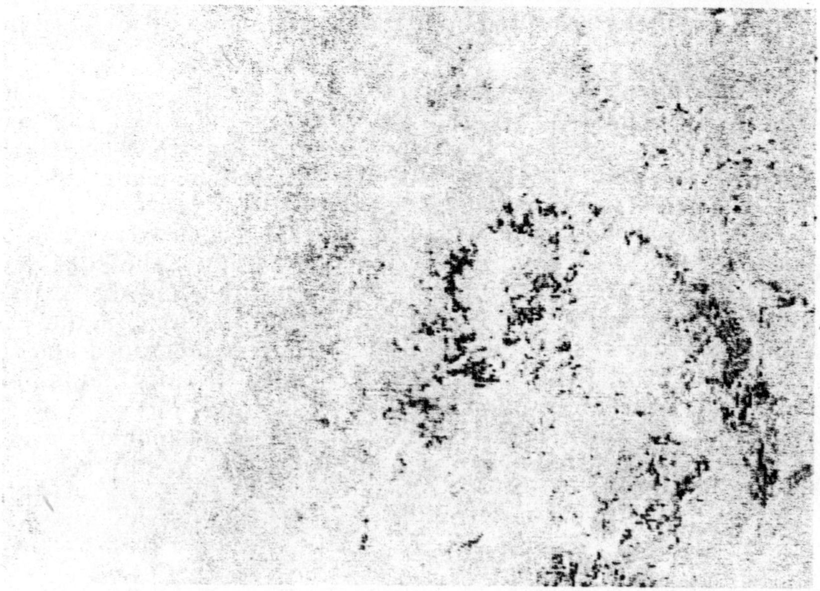

Fig. 9.1. Landsat band 5 black and white image showing tonal and spatial characteristics of hills, rocky/gravilly and flat buried pediments and associated natural resources

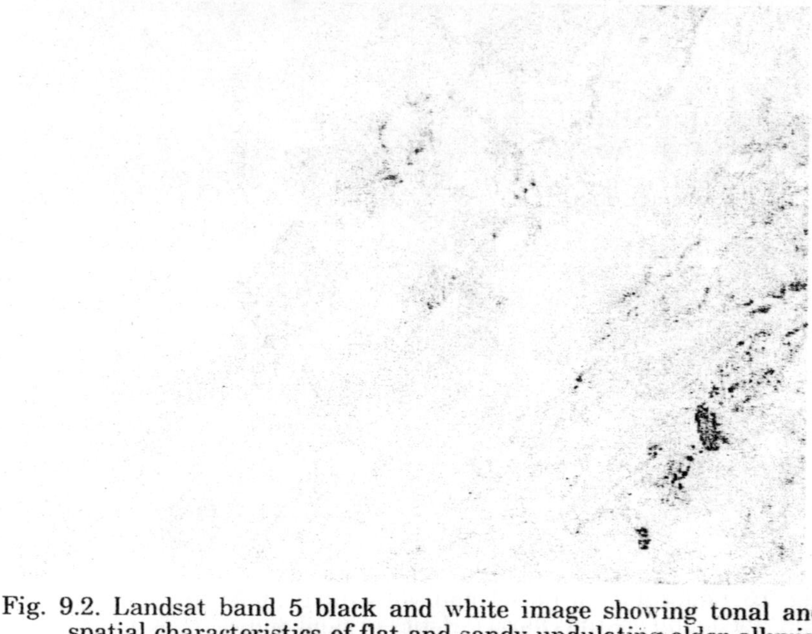

Fig. 9.2. Landsat band 5 black and white image showing tonal and spatial characteristics of flat and sandy undulating older alluvial plains, younger alluvial plains, gradea river beds, saline depressions and associated natural resources

this image (Fig. 9.3). The saline areas and river beds due to similar reflectance appear almost alike on the Landsat bands 5 and 7 images but they could be separated by distinct variations in their shapes and patterns.

By employing aerial and space remote sensing techniques and through random field checks, the State of Rajasthan has been divided into four geomorphic regions viz. the Rajasthan desert, the Aravalli mountainous region, the east Rajasthan plains and the south-eastern plateau region (Singh *et al.* 1991). Based on the tonal variations and shape, size and pattern, each geomorphic region has been further divided into major landform units. The Rajasthan desert has been divided into nine landform units and the Aravalli mountainous region has four major landform units. In the east Rajasthan plains, six landform units have been identified and the south-eastern plateau region has two major landform units. The significant variations in morphological features, slope, drainage characteristics and moisture status have caused the tonal variations in these landform units varying from white, grey to dark grey and dark grey to dark.

Fig. 9.3. Landsat band 7 black and white image showing forms, patterns, tonal and spatial characteristics of sand dunes

The small scale multispectral Landsat imagery has been found useful to study the coastal geomorphology of the western Maharashtra. Through this study, a number of landform units such as the units of structural origin, units of denudational origin, units of fluvial origin and units of marine origin have been differentiated (Sharma, 1980). Simultaneous interpretation of all the four Landsat bands has brought out that most of the near and off-shore features like active marine accretion zone, beach bars, traces of submerged off-shore bars and stream channels can be distinguished with a remarkable clarity on bands 4 and 5. Landsat band 7 in combination with band 6 found useful to distinct inland features like lateritic capping, stream courses, joint and fractures and headward erosion zone, etc.

The geomorphological map covering about 10,480 sq. km area of Tripura State has been prepared by visual interpretation of Landsat imagery (Bhattacharya 1980). In this area, seven geomorphic units viz. : (i) zone of resistant structural hills, (ii) low lying moderately dissected strike ridges, (iii) highly dissected low lying hills, (iv) denudational hills, (v) residual sandstone hills, (vi) undulating plains with low mounds and flood plains of the rivers have been identified and mapped. The salient flood plain features like natural rivers, point bars and abandoned channels could be mapped from Landsat imagery.

Analysis of the aerial photographs enabled to identify and to map eleven landform units and their characteristics in Rajkot District, Gujarat State (Murthy *et al.* 1980). The landform units interpreted and mapped are, hills, table lands, pediment plain, mud flat tidal inlet areas, stony gravelly waste, *rann* (dry salt land), *rann* (salt waste wetland), saline and sandy mounds, sandy bars and island bar. The problematic areas associated with landforms could also be identified and mapped for conservation of resources and maintenance of ecological balance in these areas.

Geomorphic mapping of a part of the fluvial and coastal belt of Andhra Pradesh including Krishna Delta was carried out by using Landsat imagery. Seven landform units with their geomorphic characteristics were demarcated and mapped (Rakshit, 1980). In the aeolian terrain of the Thar Desert four major landform units such as : alluvial plains, stable dune plains, marginal zone of active erosion and active dune plains were demarcated and mapped. Based on shape, size and pattern, dune features were identified. In the Vindhyan Decan Trap, four major landforms viz. : structural hills, denudational hills, low lying surfaces and recent fluvial plains have been identified. The morphological variations based on spectral reflectance were evaluated.

Based on the interpretation of Landsat imagery, the geomorphological mapping of Mizoram State was done and seven landform units were identified (NRSA, 1979). The landform units are: longitudinal valleys and low ridges, strikes parallel ridges and alternating valleys, structural hills and ridges and valley province, parallel mountain ranges, structurally controlled and steep valleys, mountain ranges continuous and regularly disposed, mountain ranges irregularly disposed and discontinuous and sub-parallel mountain ranges and steep valleys. The associated geomorphic features and hazards were also evaluated from Landsat imagery.

DIGITAL ANALYSIS OF LANDSAT DATA FOR LANDFORMS CLASSIFICATION

Digital analysis of Landsat imagery in the form of Computer Compatible Tapes (CCTs) has been found useful for the investigation of landforms of the middle Luni basin, western Rajasthan (Singh and Shankarnarayan, 1982). Five sub-scenes were digitally enhanced and classified in different themes such as dunes and swales, younger alluvial plains and flat buried pediments, water bodies and medium to heavy textured alluvial plains and saline alluvial plains. The spectral variations within the landforms could be detected which are not possible from visual interpretation.

Digitally edge-enhanced IRS-LISS II and Landsat TM False Colour Composites (FCCs) enabled to identify major landforms viz., hills, obstacle dunes, rocky/gravelly pediments, flat older alluvial plains, saline alluvial plains, saline depressions, younger alluvial plains and river beds in the southern part of Jodhpur District, western Rajasthan. The boundaries of these landforms were very sharp on these products which could be easily delineated and mapped (Kar *et al.*, 1989).

Soils

The soils of different types occur under various morphoclimatic zones of India. The systematic mapping of these soils is an essential pre-requisite for agricultural development. Soil mapping carried out by the soil scientists through conventional methods has been found costly, time consuming and less accurate. Remote sensing techniques based on the physiographic or geomorphic concept have been proved economic, faster and accurate in soil surveys and mapping at three levels; reconnaissance, semi-detailed and detailed. The reconnaissance, and semi-detailed soil surveys are designed to provide generalised information for low intensity agricultural and rangeland uses. The information generated by detailed soil surveys could be utilised for extensive land use and land development planning. Visual and computer analysis remote sensing techniques have been often superior to conventional soil surveys. Aerial photographs have been used for small scale mapping and the results have been used in soil surveys for soil boundary detection and visual perception of tonal qualities associated with spatial patterns. Landsat data have been found useful in identification and mapping of soils due to their spatial and spectral properties. In recent years, more systematic and detailed studies in soil classification and mapping have been done by using IRS-LISS II, III and PAN, SPOT and Landsat TM data.

The results of the studies conducted on soil surveys and mapping in different parts of the country by using remote sensing techniques in conjuction with ground truth have been reviewed.

Classification and Mapping of Soils

In the arid environment of Rajasthan, the use of aerial and satellite remote sensing techniques has been made in the classification and mapping of soils in association with landforms. Abichandani (1965) has discussed the use of aerial photo-interpretation in the detailed reconnaissance soil survey of the central Luni basin in conjuction with landforms and field surveys. Based on the morphological characteristics of landforms, broad soil classes such as sandy soils of sand dunes, medium to coarse textured soils of older alluvial plains, coarse textured soils of piedmont plains, coarse to medium textured soils of younger alluvial plains and halomorphic medium to fine textured soils of depressional areas have been identified in the basin. These soils appear in white to light grey and dark grey to dark tone depending on their texture and moisture status. The difficulty was encountered in demarcating soil boundary in the older alluvial and recent flood plains due to the absence of slope and other recognisable features. The soil boundaries, however, were delineated with the help of distinct surface ground features, vegetation characters and tonal variations.

It has been inferred from this study that the use of aerial photographs offers considerable advantages over conventional ground methods in terms of accuracy, time and cost for soil survey and mapping. In detailed reconnaissance soil survey, areas where greater variability is observed can be easily located and demarcated on aerial photographs.

Small scale soil mapping of Jodhpur District by using aerial photographs and ground truth was done and soils of different series were identified (Dhir, 1974). Sand to loamy sand soils of Chirai Series appear on aerial photographs in light grey tone with fine mottles and dark dots. Medium textured loamy sand to sandy loam soils of Pipar Series and heavy textured loamy soils of Bhopalgarh Series appear in dark grey and dark tone, respectively on air photos. The results of this study have been summarised in Table 9.1.

Landsat imagery of bands 5 and 7 and aerial photographs have been used for the identification and mapping of salt affected soils in south-eastern arid Rajasthan (Kolarkar *et al.*, 1980). They found that non-saline soils due to the presence of good vegetation and absence of salinity appear in dark grey tone on Landsat band 5 image (Fig. 9.2). The saline soils in the absence of vegetation show higher reflectance and appear in light or greyish white tone on the image of above Landsat band. Soils associated with sand dunes and sandy plains exhibit white to grey and dark grey tone on Landsat band 5 image.

By using systematic aerial photo-interpretation technique, the physiographic units and associated soils were studied in southern part of Haryana (Sangwan and Garalapuri, 1980). The level alluvial plain constitutes Typic Camborthids and soils associated with interdunal plain are coarse loamy, Typic Ustipsamments and Typic Ustifluvents. The sand dunes and sandy plains comprise Typic Ustipsamments and Typic Cambrothids.

Murthy *et al.* (1980) have established soil-physiographic relationship in Bundelkhand region of Madhya Pradesh by using aerial photo-interpretation techniques. Shallow Lithic Ustorthents soils are associated with hills and hill ridges. Loamy to fine texture Lithic Vertic/Ustochrepts occur on plateaus. Soils of pediments are mainly shallow to moderately deep with coarse loamy Typic Ustorthents. Soils of piedmont plains are deep fine Entic Chromusterts. Soils of nearly level alluvial plains are non-calcareous and calcareous Entic Chromusterts while dissected alluvial plains have Typic Ustochrepts.

An attempt has been made by Pujhari *et al.* (1980) to make use of satellite imagery for delineation of soil-scapes boundaries in the proposed command of the lower Suktel Irrigation Project of Bolangir District, Orissa. The study indicated that soil-scapes interpreted individually utilizing black and white imagery of bands 4,5,6 and 7 can provide suitable soil-scape map which is no way inferior than conventional map prepared by *Grid Method*.

Dwivedi (1985) by using black and white prints of Landsat TM bands 2, 3, 4, 5 and 7 of 500,000 scale has mapped soil resources of Karimnagar District, Andhra Pradesh in association with physiographic units. Landsat TM data with better spatial, radiometric and

Table 9.1. Photo-characteristics of some soil mapping units

	Photo-characteristics of soils	Associated features	Soil mapping units
1.	Somewhat darker tone and less rough texture than those of Chirai Series	Less mottling associated with scrub vegetation and better demarcated field boundaries than those in chirai Series	Pal series or soils with brown, sandy to loamy sand surface and some what heavier sub-soil
2.	Dark grey tone with distinct variation associated with land use, a very smooth texture	Sparse dots of tree vegetation, very distinct field boundaries, negligible mottling	Gajsinghpura Series or greyish brown heavy textured soils from limestone
3.	Medium grey, sometimes dark grey tone, smooth texture	Above normal density of dots of tree vegetation, distinct field boundaries, few mottles	Malkosni Series or brown, medium textured soil from sandstone
4.	Heterogeneous tone ranging from light to dark grey with irregular white patches in between and smooth texture	Normal density of tree dots, white patches generally do not show field boundaries	Malkosni Series or greyish brown medium texture, strongly alkaline, moderately saline soils, Barren spots saline from the surface
5.	Light grey tone with generally smooth texture	Sparse tree dots, less distinct field boundaries, proportion of area under cultivation appreciably below average	Kolu Series or light brown, moderately deep. Loamy sand soils followed at 40-60 cm depth by a thick, indurate carbonate concretionary layer
6.	Pattern of very light grey and slightly darker tone occurring in crescentic, circular or irregular forms	Sparse tree dots, cultivation confined to pockets of darker grey tones. Area under cultivation 20 to 50% only	Bap Series or moderatelly deep browm light textured soils with thick compact, cobbly strata at 40-60cm depth in association with exposed boulder beds

Source : Dhir (1974)

spectral resolution provides better contrast amongst various landscape units over MSS data, particularly bands 5 and 7 data. Dominant soils identified in association with landforms such as hills with and without vegetation and upper and lower pediments with grey tone and smooth texture are Udic Haplustabls, Lithic Ustorthents and Typic Ustochrepts.

The identification, delineation and mapping of the soils of Rajasthan and surrounding regions of India was done by using Landsat false colour composite mosaic, USDA soil geography map and ground truth (Shankarnarayan and Singh, 1979). The soil boundaries from USDA soil map when overlaid on the Landsat mosaic showed a similarity with the landform features on the mosaic. However, use of the Landsat imagery permitted refining the soil boundaries delineated from the USDA soils map for greater accuracy, at soil association level. Further scrutiny and tonal and textural matching in the mosaic led to redesignating some soils with similar spectral characteristics. For example, the regur soils (Typic Chromusterts) north of Bhilwara were earlier mapped as Typic Solorthids from alluvium on weakly dissected terraces, but on close scruntiny it was found to perfectly match with the tonal and textural character of soils delineated as regur (Typic Chromusterts) from basaltic rock east of Mandasur. Thus the Landsat false colour composite mosaic enabled efficient regrouping of soils having similar tonal and textural characteristics. Associating with landform features and analyzing the contrasting tonal and textural variations, it has been possible to distinguish twenty two soils in the region. The salient characteristics of soils together with their ratings, are presented in Table 9.2. It is significant to point out that the dominant soils occurring in this region are the Alfisols and Aridisols which are intimately associated with the rocky gravelly and buried pediments. The sand dunes and sandy plains obviously occupy large areas of the western part composed of single taxonomic unit Torripsamments, which could not be classified further to the sub-group level.

Landsat MSS data in the form of black and white imagery were used to generate soil map of Punjab in association with physiographic units (Sehgal *et al.* 1985). In the arid zone, Typic Camborthids and Typic Calciorthids are dominant soils in upper terraces. Soils associated with sand dunes are Torripsamments (Typic, Thapto Camborthidi). In the dunal areas of Ustic zone, Typic Ustipsamments and Psamments Ustichrepts are dominant soil formations. The soils in flood plain are highly stratified and qualify

Table 9.2. Landforms and associated soil characteristics in Rajasthan and surrounding regions

S. No.	Landforms	Classification and description of soils	Precipitation (mm)	Texture	Depth	Internal soil drainage	Salinity	Slope	Water holding capacity	Soil rating: Texture	Soil rating: Depth	Soil rating: Drainage	Soil rating: Salinity	Soil rating: Slope
1	2	3	4	5	6	7	8	9	10	11	12	13	14	15
1	Mountains	Entisols-Ustorthents Lithic (LM). Lithosols on mountains of arid and semi-arid areas	630-750	vc	vsh	e	0	m	vl	4	4	4	1	4
		Entisols-Ustorthents Lithic (LS). Steep Lithosols of arid regions	500-630	vc	vsh	e		m	vl	4	4	4	1	4
2.	Rocky/gravely and buried pediments	Alfisol - Paleustalf Typic (RC H/C-L). Reddish chestunt soils from consolidated materials on hilly terrain	630	vm	sh	se	0	h	1	2	3	2	1	4
		Alfisol-Rhodustalf Lithic (RB R/G)	500-630	vm	sh	se	0	r-h	1	2	3	2	1	4
		Alfisol-Rhodustalf Lithic (RB R/C). Reddish brown soils from consolidated rocks in hilly terrain	630	vm	sh	se	0	r-h	1	2	3	2	1	3
		Aridisol-Camborthids (RDg). Gravelly red desert soils	300-400	vc	md	se	0	gr-r	1	4	2	2	1	2
		Aridisol-Camborthids (RDs). Sandy red desert soils	250-300	vc	md	se	0	gr-r	1	4	2	2	1	2-3

Table 9.2. contd...

Table 9.2 contd . . .

1	2	3	4	5	6	7	8	9	10	11	12	13	14	15
		Aridisol-Solorthids Typic (SR T/A). Sierozems from alluvium on weakly dissected terraces	500-630	vm	d	w	0	nl	h	1	1	1	1	1
3.	Rocky/gravelly and buried plateaus	Alfisol-Durixeralfs Typic (NB R/SD). Non-calcareous brown soils from sandstone on rolling terrain	500-630	vc	sh	se	0	nl-h	1	3	3	2	1	3
		Alfisol - Durixeralfs Typic (NB H/SD). Non-calcareous brown soil is from sand-stone on hilly terrain	630-750	vc	sh	se	0	h	1	3	3	2	1	4
4.	Light to medium textured older alluvial plains	Aridisols- Solorthids Typic (SR T/A). Sierozems from alluvium on weakly dissected terraces	500-630	vm	d	w	0	nl	h	1	1	1	1	1
		Aridisols - Solarthids Aquollic - (SRA). Sierozems from alluvium												
5.	Medium textured older alluvial plains	Aridisols - Solorthids Typic (SR T/A). Sierozems from alluvium on weakly dissected terraces	500-630	vm	d	w	0	nl	h	1	1	1	1	1
6.	Heavy textured older alluvial plains	Vertisols - Chromosterts (RG R/B). Regur soils from basic igneous rocks on undulating to rolling terrain	630-880	vf	d	p	0	r	h	3	1	3	1	2
		Aridisols - Solorthids Aquollic - (SKA). Solonchak soils from recent alluvium	630-880	vn	d	1	m-se	nl	m-h	1	1	3	3	1

Table 9.2. contd...

1	2	3	4	5	6	7	8	9	10	11	12	13	14	15
7.	Sand dunes and sandy plains	Entisols - Torripsamments Typic - (FSD). Fine sand to loamy fine sand	100-350	cs	d	1	0	h	vl	4	1	4	2	2
8.	Younger alluvial plains	Entisols - Hydraquents (AH). Hydromorphic alluvial soil	750 - 880	vm	d	p	s	nl	m	2	1	3	1	1
		Entisols - Ustifluvents (ACN). Coarse textured clacareous alluvial soil	750-880	ls	d	sp	sl	nl	h-m	3	1	2	2	1
		Entisols-Ustifluvents (AN). Non-calcareous alluvial soil	750-850	vm	d	w	m	hl	h	2	1	1	2	1
		Alfisol-Paleustalf Typic (RCA). Reddish chestnut soils from alluvium		vm	d	w-sp	0	gr	h	1	1	1	1	2
		Alfisols - Paleustalf Typic (RC T/A). Reddish chestnut soil from alluvium on weakly dissected terraces	400-880	vm	d	w	0	gr	h	1	1	1	1	2
		Alfisols-Paleustalf Typic (RC DT/A). Reddish chestnut soil from alluvium on disected terraces	750-880	vm	d	w-sp	0	nl-r	m	2	1	1	1	1-3
9.	Rann (Playa)	Aridisol - Solorthids Aquollic (SM). Salt water marsh	200-250	vf	d	v-p	sc-vs	nl	h	3	1	3	3	1

Source : Shankarnarayan and Singh (1979).

for Ustifluvents (Aquic/Typic). In the piedmont plains and hills, dominant soils are Ustipsamments (Fluventic and Ustochrepts Typic/Udic).

The data available from Salyut - 7 space mission have been evaluated for soil mapping in two areas with different geological and geomorphological formations located in Bidar District of Karnataka and Zaharabad area of Medak District of Andhra Pradesh (Rao and Venkataratnam, 1985). The results indicate that for soil-scape boundary delineation, Terra data is more useful because of high spatial and spectral resolution. The FCC of MKF-6M gave better classification of landcover, eroded areas, major gelological formations and coastal landforms. Kate- 140 is found highly suitable for physiographic analysis even in the coastal and deltaic areas where relief variations are very low. The conjunctive use of MKF-6M and Kate-140 data give better results as compared to Landsat in soil-scape boundary delineation. Among the various bands of MKF-6M, band 3 (0.79-0.89 m) or 4 (0.64-0.68 m) in combination with band 6 (0.58-0.62 m) which are comparable with Landsat bands 5 and 7, are suitable for soil-scape boundary delineation.

DIGITAL ANALYSIS OF SOILS CLASSIFICATION

The supervised computer classification of Landsat TM sub-scene comprising of south-eastern part of Jodhpur District in conjunction with ground truth enabled to identify 12 soil classes. Supervised computer classification of IRS-LISS II sub- scene could bring out only 8 soil classes due to slightly poor resolution. Among these soil classes, Pipar Series occupy the largest area followed by Chirai Series (Singh *et al.* 1989).

Based on the analysis of Landsat data in the form of Computer Compatible Tapes (CCTs) of seven scenes, soil classification and mapping of Haryana State in conjunction with physiography and ground truth was done by using Multispectral Data Analysis System (MDAS). The soils classification upto associations of sub-groups level was possible due to inherent limitations of Landsat data such as poor resolution and small scale. The results of computer soil classification and mapping of one Landsat scene 158-039 representing all the physiographic units and soil types of Haryana State are presented in Table 9.3 (NRSA, 1979).

Table 9.3. Soil classification for Landsat scene 158-039

Colour	Soil association	Physiography	Present landuse or crops grown	Remark/Limitations
1	2	3	4	5
Dark red	Typic Ustochrepts Typic Ustorthents Typic Ustifluvents Fluventic Ustochrepts	Alluvial plains nearly level	Insensively cultivated land	Sandy loam or finer texture soils, give higher reflectance values
Bright red	Typic Ustorthents Typic Ustochrepts	Alluvial plains nearly level to very gently sloping	Cultivated area	Loamy sand to sandy loam soils, give higher reflectance values than the above soils, well drained
Yellow	Typic Ustipasmments Typic Ustorthents	Gently to moderately sloping sandy areas in the transition zone between Siwaliks and plains, yellow colour also denotes Typic Udipsamments and Typic Udorthents	Cultivated if irrigation facilities are available Millet grown during rainy season	Sandy to sandy loam soils, give higher reflectance values than the above soils, well to excessively drained
White	Typic Ustipsamments Typic Udipsamments Eithic Udipsamments	Dissected rolling plains of North Haryana below Siwalik hills. River sand also seen in white colour.	Cultivated if irrigation facilities are available	Sandy soils, give higher reflectance values than the above soil associations, excessively drained

Table 9.3. contd...

Table 9.3 contd . . .

1	2	3	4	5
Purple	Typic Natrustalfs Aquic Natrustalfs	Strongly saline/ alkaline nearly level plains	Barren land	Loamy sand to loam soils, poor soil structure, imperfectly drained, give very high reflectance values because of white encrustations of salts, salinity and sodium hazards
Light pink	Natric/Typic Camborthids Typic Calciorthids Typic Natrustalfs Typic Natrargids	Saline/alkali land on nearly level plains	Barren land with grass species resistant to salinity	Loamy sand to loam soils, imperfectly drained, give less reflectance values than highly saline soils. Salinity/sodium hazards present
Green	Hills/Forests/Scrubs	—	—	—
Blue	Tanks/Rivers	—	—	—

Source : NRSA (1979)

Criteria used for rating soil characteristics in Table 9.2 is given below

A. Texture

vc - Variable, mostly coarse
vm - Variable, mostly medium
vf - Variable, mostly fine
cs - Coarse sand
ls - Loamy sand

B. Depth

vsh - Very shallow <10 inches
sh - Shallow 10-20 inches
md - Moderately deep 20-30 inches
d - Deep >36 inches

C. Internal soil drainage

vp - Very poorly drained

p - Poorly drained

sp - Somewhat poorly drained

mw - Moderately well drained

w - Well drained

se - Somewhat excessively drained

D. Salinity (mmhos/cm)

o - None (<2)
sl - Slight (2-4)
m - Moderate (4-8)
se - Severe (8-16)
vs - Very severe (>16)

E. Slope

nl - Level gently slopping (0.5%)
gr - Gently rolling (5-10)
r - Rolling (10-15%)
h - Hilly, steep (15-30%)
m - Mountainous (>30%)

F. Water holding capacity

vl - Very low (<3.2 inches in root zone)
l - Low (3.2 - 4.8 inches in root zone)

m - Medium (4.8-6.4 inches in root zone)
h - High (> 6.4 inches in root zone)

FOREST AND VEGETATION

In India, the forests constitute 75 million hectares (22.8%) of total geographical area. The National Forest Policy lays down that not less than 33% of the land surface of the country should be kept under forests for the welfare of the human beings and economic development of the Nation. The existing forest cover is also depleting due to the increasing biotic activities like deforestation, overgrazing and cultivation of marginal lands. In order to check the further deterioration of forests and vegetation by proper planning and management and to bring more land under them, the systematic survey, classification and mapping of forests and vegetation are an essential pre-requisite. The conventional techniques used for preparing the forest inventory upto recent times have been found tedius, time consuming, costly and less accurate, particularly for inaccessible areas. Aerial and space remote sensing techniques have been proved economic, faster and accurate for classification and mapping of forests and vegetation. The use of aerial photo-interpretation for forest resources survey and mapping over large areas has, however, began with the establishment of Pre-investment Survey of Forest Resources Organisation in 1965 (now reconstituted as the Forest Survey of India). Since then, this technique is being used by various Government and Non-govermnet departments for surveying and mapping of forests and vegetation.

With the launch of U.S. Landsat-1 (initially designated as ERTS-1) in 1972, the satellite remote sensing has ushered in a revolutionary change in the surveying and mapping of forests. The information acquired from satellite is near real time and absolutely relevant for planning the development activities. Based on the visual and digital analysis of Landsat MSS and TM data, surveying and mapping of forests and vegetation have been done in different parts of the country. The data generated by French satellite SPOT and Indian Remote Sensing Satellites (IRS-1, A,B,C,D) have increased the potentiality of the satellite remote sensing in forest survey and mapping.

The work done through remote sensing techniques on forests and vegetation by various organisations in the country has been reviewed.

Classification and Mapping of Forests and Vegetation

The use of small, medium and large scale black and white aerial photographs has been extensively used by the Pre-investment

Survey of Forest Resources Organisation for the survey, classification and mapping of forests in different parts of the Country (Tomar, 1972). The results of these studies have indicated that on small scale photographs (1 : 40,000-1 : 60,000); it is difficult to identify individual species except large gregarious patches or plantation areas of one species or separation of coniferous stands from broad leaved in some areas. However, the forests into broad physiognomic group such as dry deciduous, moist deciduous and evergreen forest could be identified and mapped. On medium scale photographs (1 : 20,000-1 : 40,000), various species in coniferous forest of Himalaya and in mixed deciduous forest pure or gregarious patches of teak, sal, bamboo, etc. have been identified and mapped. On large scale photographs (1 : 10,000- 1 : 20,000) taken in suitable season, some species in the deciduous or evergreen forest could be identified and mapped. The identification of gregarious stand of teak, sal or other species, occurrence of bamboos, reeds and young forestry plantations in conjunction with ground truth could be easily identified and mapped.

Estimation of height, crown density and crown diameter has been done by aerial photographs which is an essential pre-requisite for forest inventory and management. In the mixed deciduous forests of Bastar, stand height measurements were carried out on various scales of photographs. The height measurements on 1:15,000-1:20,000 scale aerial photographs is fairly reliable in even-aged trees occurring on flat to gently undulating topography. A height class of 5m interval can be safely used for stratification of the stands. Estimation of crown density and crown diameter can be quickly made on aerial photographs with an acceptable degree of accuracy. The forests can be easily stratified into 20% density classes on medium scale photographs. On small scale photographs three broad density classes such as less than 40%, 40-70% and more than 70% may be conveniently adopted. In the regular teak forest of Allapalli (Chanda) 5 m crown diameter class on 1 : 10,000 scale photographs attempted to prepare aerial volume table.

Landsat visual and digital analysis techniques have been successfully used for operational projects by NRSA (1985) for surveying, mapping and monitoring forest features and phenomenon in different parts of the Country. The salient findings of the operational projects are presented in Table 9.4.

Forest thematic map based on Terra data and ground truth has been prepared for Dangs forest, south Gujarat region. The forest areas have been mapped into virgin forests, closed forests, degraded

Table 9.4. Remote sensing project in forest/vegetation mapping done by NRSA

S.N.	Region	Area covered (km^2) (approximately)	Data used	Remote sensing data utilised/Method of analysis and interpretation	Output	User agency
1	2	3	4	5	6	7
1.	Andhra Pradesh	70,000	Landsat scene 153- 048 & 050	Computer Compatible Tapes (CCTs), False Colour Composites (FCCs), B&W imageries of band 5 and 7, Computer analysis and interpretation	Colour coded forestry maps delineating different forest types (includings degraded) and forest vegetation on 1 : 250,000 scale	Government of Andhra Pradesh
2.	Basthar Distrct Madhya Pradesh		Landsat scenes 152-047; 153-046 to 048 (4 scenes)	B & W imageries; visual interpretation	Line map delineating different forest types (1 : 1 M scale)	Madhya Pradesh Forest Development Corporation Ltd. Jagadalpur
3.	Bundelkhand and Adjoining Areas Uttar Pradesh	41,283	Landsat scenes 155-042; 156-042; 156-043 (3 scenes)	CCTs, FCCs, B & W imageries of band 5 and 7; computer analysis and interpretation	Forest type maps (1:250,000 scale) including forest classes like miscellaneous, deciduous and degraded forests area calculated under each category	Council of Science and Technology Govt. of Uttar Pradesh
4.	Lower Barak Rever	6,962	Landsat scenes 146-043 (1 scene)	CCTs, FCCs, B & W imageries of Band 5 and 7; Computer analysis and interpretation	Forest type maps including mapping of shifting cultivation and degraded forest areas on 1 : 250,000 scale	North-Eastern Council (NEC) Shillong

Table 9.4. contd...

Table 9.4 contd . . .

1	2	3	4	5	6	7
5.	Nagaland	14,500	Landsat scene 145-042 (1 scene)	CCTs, FCCs, B & W imageries of Band 5 and 7; Computer analysis and interpretation	Forest type maps including mapping of shifting cultivation and degraded forest areas on 1 : 250,000 scale	NEC Shillong
6.	Tripura	10,024	Landsat scenes 146-043; 147-044 (2 scenes)	CCTs, FCCs, B & W imageries of Band 5 and 7; Computer analysis and interpretation	Forest type maps including maping of shifting cultivation which includes different stages of shifting cultivation area calculated under each category and degraded forest areas on 1: 250,000 scale	NEC Shillong
7.	Mizoram	17,675	Landsat scenes 145-044; 146-043; 146-044 (3 scenes)	CCTs, FCCs, B & W imageries of Band 5 and 7; Computer analysis and interpretation	Forest type maps including maping of shifting cultivation and degraded forest areas on 1 : 250,000 scale	NEC Shillong
8.	Arunachal Pradesh	44,700	Landsat scenes 144-041;145-041; 146-041; 147- 041 (4 scenes)	CCTs, FCCs, B & W imageries of Band 5 and 7; Computer analysis and interpretation	Forest type maps including mapping of shifting cultivation and degraded forest areas including abandoned *Jhum* area on 1 : 250,000 scale	NEC Shillong
9.	Meghalaya	22.490	Landsat scenes 146-042; 147 - 042; 148-042; 148-042 (4 scenes)	CCTs, FCCs, B & W imageries of Band 5 and 7; Computer analysis and interpretation	Forest type maps including mapping of shifting cultivation and degraded forest areas including abandoned *Jhum* area on 1 : 250,000 scale	NEC Shillong

Table 9.4. contd...

Table 9.4 contd . . .

1	2	3	4	5	6	7
10.	Mikir & North Cachar Hills Assam	—	Landsat scene 146-042 (1 scene)	CCTs, FCCs, B & W imageries of Band 5 and 7; Computer analysis and interpretation	Forest type maps including mapping of shifting cultivation and degraded forest areas on 1 : 250,000 scale	NEC Shillong
11.	Idukki Kerala	8,839	Landsat scenes 150-053, 156-053 (2 scenes) and Panchromatic B & W aerial photos on 1 : 25,000 scale	CCTs, FCCs, B & W imageries of Band 5 and 7; Computer analysis and interpretation	Forest type mapping and change detection using data of two periods on 1: 250,000 scale	Department of Environment Govt. of India
12.	Silent valley Kerala	About 10,000	Landsat scense 155-052 (1 scene) and Panchromatic B & W aerial photos on 1 : 25,000 scale	CCTs, FCCs, B & W imageries of Band 5 and 7; Computer analysis and interpretation	Forest type mapping including plantations on 1 : 250,000 scale	Department of Environment Govt. of India
13.	Nationwide Forest cover	32,87,790	(190 scenes)	CCTs, FCCs, B & W imageries of Band 5 and 7; Computer analysis and interpretation	Broad forest cover mapping including close open/ degraded and man-grove forest of two periods on 1 : 1000,000 scale	Department of Space Govt. of India

Source : NRSA (1985)

forests/young plantations and forest blank areas (Aggarwal and Jadhav, 1985). A comparative study of forest stock mapping of Dangs forest done by using aerial, Landsat, Kate-140 and MKF-6 data on same scale (1 : 63,360) revealed that high resolution Terra data are useful in mapping forest areas at large scale upto 1 : 50,000 MKF-6 data, especially band 4 (640-680 nm) is the best suitable band for vegetation mapping. Landsat data is useful for strategic planning whereas MKF-6 data can provide base-line information for operational planning and working plan follow up.

Based on the visual interpretation of Landsat False Colour Composite (FCC), following 9 categories in forests of Gorakhpur District, Uttar Pradesh have been identified and mapped (Kachhwaha, 1985).

1. Dense Sal dominated forests
2. Medium to low density Sal dominated forests
3. Riparian vegetation mainly *Syzygium cumini* (Jamun)
4. Young plantations
5. Very young plantations (saplings) mixed with agricultural crop locally known as *Tangia*.
6. Degraded forests
7. Clear felled areas
8. Forest blanks
9. Water bodies

The major part of the forests in the study areas is occupied by Sal dominated forests. These forests appear in dark red tone with smooth texture on False Colour Composite (FCC).

The Landsat false colour composite mosaic of Rajasthan on 1:1000, 000 scale *per se* did not permit the identification of vegetation except in the hills where the more dense, natural vegetation has high infra-red reflectance and appears in bright red colour as viewed on the imagery. Thus, by and large, the interpretation of vegetation has mainly been made on the basis of associated features of landforms in conjunction with ground truth data. The following are the main vegetation types identified and mapped in Rajasthan (Shankarnarayan and Singh, 1979).

1. *Moist deciduous vegetation* : The southern and south-eastern part of the Aravalli hills support the moist deciduous vegetation comprising of *Tectona grandis, Anogeissus latifolia* and *Terminalia* spp.

2. *Semi-evergreen vegetation* : The Aravalli mountains in the vicinity of Jaipur support this vegetation which is characterized by the dominance of *Anogeissus pendula* virtually forming pure associations with understory formed of *Adathoda vesica*.

3. *Desert thorn vegetation* : On the plains, west of the Aravalli mountains, the desert thorn vegetation occurs which primarily consists of species such as *Prosopis cineraria, Capparis decidua* and *Ziziphus nummularia*. However, along river courses or in areas with high water table, *Salvadora oleoides* and *S. persica* occur. While *Acacia senegal* prevails on hills and pediments, species such as *Calligonum polygonoides* and *Leptadenia pyrotechnica* characterize the sand dunes and sandy plains.

4. *Northern tropical thorn vegetation* : This occurs on the plains east of Aravalli and mainly consists of *Acacia nilotica, A. leucophloea* which gets intermixed with *Acacia catechu* and *Butea monosperma,* particularly in areas of south Dholpur.

Digital Analysis of Landsat Data for Forest and Vegetation Classification and Range Biomass Estimation

Digital analysis of IRS-LISS II sub-scene of September, 1988 enabled to demarcate 5 broad vegetation zones in south-eastern part of Jodhpur District, western Rajasthan. The grazing lands could not be separated, perhaps due to similar spectral signatures of the adjoining crop fields (Singh *et al.,* 1989).

Digital analysis of Landsat MSS band 5 CCT was done and the range biomass was estimated on the basis of six grey levels in three different soil types of Jodhpur District (Shankarnarayan and Singh, 1983).

The average biomass yield was the highest in younger alluvial soils (556.8% kg/ha) followed by Chirai soils (401.9 kg/ha) and Pipar soils (371.2 kg/ha) to the normal limits obtained from three soils in clipping trials.

The range biomass computed in youger alluvial soils is given in Table 9.5.

It may be seen (Table 9.5) that the mean values of the number of pixels range between 4.75 to 157.50. The total area constituted by the pixels varies from 1.9 to 63.0 ha. The total biomass computed in the younger alluvial soils from different gray levels ranges between 36.1 to 35,910.0 kg and the average range biomass yield comes to 556.8 kg/ha.

Table 9.5. Range biomass at test site in younger alluvial soils from ND6 computer print out

Gray level Nos.	Sub-sample number and number of pixels				Mean	Total area (ha)	Total biomass (kg)
	1	2	3	4			
1	12	5	1	1	4.75	1.9	36.1
2	102	112	116	40	92.50	37.0	14,060.0
3	148	176	166	140	157.50	63.0	35,910.0
4	33	26	34	78	42.75	17.1	12,996.0
5	12	0	3	28	10.75	4.3	4,085.0
6	13	1	0	13	6.75	2.7	3,078.0
						126.0	70,165.1

Average yield = 556.8 kg/ha

The range biomass calculated at different sites under Piper soils is given in Table 9.6.

Table 9.6. Range biomass at test sites in pipar soils from ND6 comuter print out

Grey level Nos.	Sub-sample number and number of pixels				Mean	Total area (ha)	Total biomass (kg)
	1	2	3	4			
1	0	1	1	0	0.5	0.2	23.2
2	65	45	49	67	56.5	22.6	5243.2
3	156	156	119	187	154.75	61.9	21541.2
4	75	104	112	63	88.50	35.4	16425.6
5	4	12	36	3	13.75	5.5	3190.0
6	0	2	4	0	1.50	0.6	420.0
						126.2	46483.2

Average yield = 371.2 kg/ha

The mean of the number of pixels (Table 9.6) varies from 0.5 to 154.75 and the total area ranges between 0.2 to 61.9 ha. The average yield of the biomass worked out from the total biomass and the total area comes to 371.2 kg/ha.

The range biomass computed from ND6 values in Chirai soils is given in Table 9.7.

Table 9.7. Range biomass at test sites in chirai soils from ND6 computer print out

Grey level Nos.	Sub-sample number and number of pixels				Mean	Total area (ha)	Total biomass (kg)
	1	2	3	4			
1	1	0	0	1	0.5	0.2	26.6
2	67	30	95	94	72.5	29.0	7744.0
3	191	163	156	158	168.25	67.3	26852.7
4	57	108	59	64	72.00	28.8	15321.6
5	4	10	0	3	4.25	1.7	1130.6
						127.0	51045.4

Average yield = 401.9 kg/ha

It is observed (Table 9.7) that the mean of the number of pixals ranges between 0.5 to 168.25 and the total area constituted by the pixels varies from 0.2 to 67.3 ha. The total range biomass worked out from ND6 computer printout varies from 26.6 to 26,852.7 kg and the average yield of the biomass comes to 401.9 kg/ha.

LAND USE

In India, human and livestock population is increasing at an alaraming rate. During the period 1951-91, the population had increased by 90%. In order to meet the demand of this increasing population for food, fuel, and fodder, it is to plan the use of land in a scientific and rational manner. But to achieve this aim, systematic classification and mapping of present land use is necessary. The survey and mapping of present land use by conventional methods are costly and time consuming. By the time, such maps reach to public and planners, the information obtained by these maps becomes out of date due to subsequent changes. Remote sensing techniques, particularly from satellite platform due to synoptic view of large area and its spectral and temporal characteristics provide comparatively faster and cheaper way of collecting information about land use. During the decade, the visual and digital remote sensing techniques have been extensively used by several organisations in India for surveying, classification and mapping of land use in conjunction with ground truth. The visual interpretation method, using Landsat black and white and false colour composites is quite suitable for developing countries like India due to its simplicity and low cost. Colour infra-red photography (CIR) and multi-band aerial photography have been proved most important tool for detailed land use mapping. The landuse mapping, by using remote sensing techniques acquired a quasi-operational status in the country earlier has now achieved operational status due to high resolution products available from IRS-LISS II, III, PAN and Landsat TM sensors. The results of the studies conduced by different organisations on remote sensing of land use have been reviewed.

Classification and Mapping of Land Use

On the basis of the land use survey conducted in different districts of western Rajasthan with the help of aerial photographs, following major landuse units have been identified and mapped (Sen, 1978).

1. Cultivated lands 2. Sandy wastes 3. Rocky wastes
4. Gravelly wastes 5. Saline wastes.

Cultivated lands along with short fallows can be distinguished by regular and check board pattern. The current cultivated plots have medium to dark tone. The lands that have been cultivated two years back have lighter tone. The long fallows have light grey to white tone. Sand dunes and sandy plains fall under the category of sandy wastes. Sand dunes due to their distinct shape and pattern could be easily identified from aerial photographs. Sandy wastes with good stand of vegetation appear in dark grey to dark tone whereas bare sandy areas exhibit grey to white tone on aerial photographs. Hills and rocky/gravelly pediments have been termed as rocky/gravelly wastes. These wastelands distinctly appear in white to grey and dark grey tone on aerial photographs. Image structure is mottled and texture is medium to coarse. Saline depressions and saline alluvial plains form saline wastes. Due to distinct shape, size and white to grey tone, the saline wastes could be easily identified and mapped from aerial photographs.

In the Idukki District of Kerala, a land use project, based on multiband aerial photography was jointly conducted by Space Application Centre, Ahmedabad and Kerala State Land Use Board. Classification scheme used for mapping is shown in Table 9.8 (Sahai *et al.* 1980).

Table 9.8. Classification scheme for land use/land cover mapping

A. Agricultural area

A 1 Paddy fields

A 2 Sugarcane

A 3 Rubber

A 3.1. Freshly planted

A 3.2. Well grown

A 4 Tea

A 5 Coconut

A 6 Tapioca

A 7 Mixed crops

A 8 Miscellaneous crops

B. Forest

B 1 Eucalyptus plantations

B 2 Teak plantations

B 3 Mixed forests/miscellaneous

B 4 Reeds

B 4.1 Pure reeds

B 4.2 Mixed reeds

B 5 Forest blanks

B 6 Forest fire affected areas

C. Wasteland

C 1 Waste and barren

C 2 Rock outcrops

C 3 Eroded land (rills and gullies)

C 4 Marshy areas

D. Water bodies

D 1 Lakes

D 2 Streams

D 3 Rivers

D 4 Canals

D 5 Tanks, ponds, wells

E. Public use

E 1 Roads

E 2 Rail road

F. Unsurveyed/Unclassified

Source : Sahai *et al.* (1980)

The project resulted in preparation of 90 land use maps for the entire district on 1 : 15,000 scale each covering about 9 x 9 km area. Based on these maps, a comprehensive, *Land Use Plan for Idukki District* has been prepared by KSLUB (1984).

The multidate Landsat scenes of middle Luni basin, western Rajasthan, were used for classification and mapping of irrigated and non-irrigated land use units (Shankarnarayan and Singh, 1979). The following seven irrigated lands use units from Landsat false colour composite of January, 1977 have been delineated : (1) irrigated cropland, (2) fallow alluvial plains, (3) fallow sandy plains, (4) fallow sand dunes and interdune plains, (5) fallow buried pediments and plateau, (6) stony/gravelly/rocky waste, (7) saline waste.

The irrigated croplands are mostly concentrated in the younger alluvial plains. They appear in light to dark red colour on Landsat false colour composite image. The fallow lands appear in bright to dark green colour. The saline rocky and gravelly wastes exhibit a light tone and do not register any distinct colour. The irrigated croplands are also observable distinctly on Landsat band 5 image in dark tone.

The rainfed (non-irrigated) land use map was also prepared from the Landsat false colour composites of September, 1972 and 1973. Complementary information was derived from wet season Landsat band 5 image. It is observed that the rainfed cropland appeared in light to dark red tone on Landsat false colour composite and in dark tones on the band 5 image. Using this technique, four major non-irrigated land use units have been delineated : (1) non-irrigated croplands, (2) fallow lands, (3) stony/rocky/gravelly waste, (4) saline waste.

The false colour composite image appears eminently suitable for delineating and mapping present land use because the cultivated lands in red to dark red colour appear prominently in sharp contrast to the surrounding areas. It is not improbable that the light red to dark red tone of cultivated crops are indicative of the vigour of the crops and could be advantageously exploited for differentiation of healthy and diseased crops.

Both visual interpretation and digital processing techniques have demonstrated the capability to distinguish land use categories of USGS classification level I and level II in MSS data (Table 9.9). In certain cases, it has been possible to achieve level III of USGS classification (Sahai *et al.*, 1980). Level I categories can be identified generally at 90% accuracy or better. Level II categories can be identified at an accuracy of 80 to 85% with the exception of the categories comprising urban or built-up area.

Table 9.9. Landuse and land cover classification system for use with remote sensor data

	Level I		Level II
1.	Urban or built-up land	1.1	Residential
		1.2	Commercial and services
		1.3	Industrial
		1.4	Transportation, communications
		1.5	Industrial and commercial complexes
		1.6	Mixed urban or built-up land
2.	Agricultural land	2.1	Cropland and pasture
		2.2	Orchards, groves, vineyards, nurseries, ornamental and horticultural areas
		2.3	Confined feeding operations
		2.4	Other agricultural land
3.	Range land	3.1	Herbaceous range land
		3.2	Shrub and bush range land
		3.3	Mixed range land
4.	Forest land	4.1	Deciduous forest land
		4.2	Evergreen forest land
		4.3	Mixed forest land
5.	Water	5.1	Streams and canals
		5.2	Lakes
		5.3	Reservoirs
		5.4	Bays and estuaries
6.	Wetland	6.1	Forested wetland
		6.2	Non-forested wetland
7.	Barren land	7.1	Dry salt flats
		7.2	Beaches
		7.3	Sandy areas other than beaches
		7.4	Bare exposed rock
		7.5	Strip mines, quarries and gravel pits
		7.6	Transitional areas
		7.7	Mixed barren land
8.	Tundra	8.1	Shrub and bush tundra
		8.2	Herbaceous tundra
		8.3	Bare ground tundra
		8.4	Wet tundra
9.	Perennial snow or ice	9.1	Perennial snow fields
		9.2	Glaciers

Source: Anderson *et al.* (1976)

With the help of MDAS and visual interpretation techniques using Landsat MSS imagery, land use surveys in parts of Andhra Pradesh, Haryana, west coast region, Nagaland, Orissa, Mizoram, Tripura, Arunachal Pradesh and Uttar Pradesh covering about 45 million hectare area have been carried out (NRSA, 1983). In some cases, available aerial photographs were utilized for the studies in conjunction with gound truth. The final land use maps have been prepared on 1 : 250,000 scale or 1 : 1 million scale. In these areas, 23 land use/land cover categories were identified and divided into 5 sections such as : water, crops, forest, small grains and others. These maps have been used by State Soil Conservation Departments, Forest Departments and Soil Survey Departments for tackling various problems of the States and also for general land use planning and management.

Based on the visual interpretation of multidate False Colour Composite (FCC) of 1 : 250,000 scale, Shanware *et al.* (1985) have identified and delineated landscape facets comprising of hills, intermontane valleys, pseudo-piedmonts and aeolian plains. In relation to these facts, four landuse patterns such as : agriculture, wastelands, agro-pastoral zone and forest were mapped. The intensity, extent and distribution of different colour tones on FCC facilitated recognition of *rabi* crops (red), fallow lands (yellow), perennial grasses (light magenta), seasonal grasslands (greenish brown and yellow), scrub forest (dark greenish/brownish red) and wastelands dissected and non- dissected (greenish brown, fine to broad white streak).

Different data products including enhanced Landsat imagery have been used for mapping the land use features and terrain conditions of the Sirohi District, Rajasthan, India (Gupta and Rao, 1985). A total number of 21 features grouped under five major categories have been identified on the Landsat linear stretched false colour composite. It was observed that different features on the Landsat imagery could be mapped to different levels of land use classification. The study has indicated that linear stretched false colour composite data of Landsat is useful for mapping the land use features as compared with the raw data of FCC. It is also possible to differentiate and map the fallow land, cropped land, wastelands, topography, towns and roads sizes of approximately 125 acres. The mapability of these features with simulated LISS-II imagery of Indian Remote Sensing Satellite was 94%. The newly afforested areas that were not mapped earlier in the district, forest maps could be mapped using Landsat imagery. The details of the features mapped clearly on different data products are presented in Table 9.10.

Table 9.10. Details of the features mapped clearly on different data products

Data product	Landsat	Features
Raw band	4	x
	5	Drainage, crops, soils, rock types and structures, roads
	6	Stony waste
	7	Water, topography
Linear stretched data band	4	Drainage
	5	Drainage, barren lands. soils, rock structure
	6	Barren lands, soils, rock types, roads
	7	Irrigated fields, barren lands, topography, rock structures
Rative imagery	4/6	Drainage, crops, fallow lands,scrub lands
	5/7	Drainage, crops, fallow lands
	4/7	Crops
	5/6	x
High band poss enhancement	4	Drainage, topography, rock structures
	5	Drainage
	6	Stony waste
	7	Stony waste, topography, rock structures
Principal component	4	Stony waste, rock types, topography, rock structures
	5	Rock exposures
Histogram equalised	5	Abandoned channels
	6	Drainage, abandoned channels, crops, forest, fallow lands, scrub lands, soils, stony waste, rock exposures, topography
	7	x
Edge enhancement	4	Drainage, rock exposures, topography
	5	Topography
	6	Forests, stony waste, topography
	7	Forests, stony waste, topography
FCC raw		Drainage, water, abandoned channels, crops, fallow lands, black soils, stony waste, topography, rock structures, urban lands, roads
Linear stretched		All the above plus irrigated fields, scrub lands, barren lands, rock exposures
IRS imagery simulated		
LISS-I		Rock types and all above except abandoned channels and barren lands
LISS-II		All above plus abandoned channels

Source : Gupta and Rao (1985)

Digital Analysis of Landsat Data for Land Use Classification

Digital analysis of Landsat TM data of April, 1987 helped to classify south-eastern part of Jodhpur District into seven landuse classes. They are sand dunes, saline waste, present cropped area, hills, fallow land, pastures and gravelly wastes. At some places gravelly wastes are misclassified, IRS-LISS II in the form of Computer Compatible Tape (CCT) for the months of September and December were also digitally analysed and rainfed and irrigated land use classes were identified. Some landuse classes such as pastures and gravelly wastes could not be separated and similarly river bed and saline *rann* (Singh *et al.,* 1989).

WATER RESOURCES

Water is the most versatile and vital resource for the existence of all forms of life such as human and livestock consumption and for industrial and agricultural activities. In order to meet the increasing demand of water, systematic studies on the identification, classification and mapping of surface and ground water resources are of paramount importance. The inventory of water resources through conventional methods is costly and time consuming. Moreover, in inaccessible areas, it is rather difficult to locate the potential surface and ground water areas for their harvesting, exploration and management. Remote sensing techniques, aerial and space have been found economic and cheaper for evaluation and development of water resources. Studies conducted in different parts of the country, have revealed that surface water in the form of rivers, streams, tanks and reservoirs could be easily and quickly identified and mapped from aerial photographs, IRS and Landsat black and white and false colour composite, particularly from Landsat band 7 image. Aerial photographs due to their three dimensional characteristics enable to delineate the boundaries of watersheds accurately and to compute their morphometric characteristics. The data thus generated will help to identify priority watersheds for their development and management.

Investigations conducted on the exploration, location and mapping of ground water in different parts of the country have revealed that ground water bearing formations are, by and large, covered with sand and alluvium. The location and mapping of these formations by geophysical and electrical resistivity methods are costly and time consuming and do not yield meaningful results. Like surface water, the direct interpretation of ground water from aerial photographs and Landsat imagery is not possible but it could be located by indirect methods which are of geomorphological nature. The location and mapping of ground water, using remote sensing techniques in conjunction with geomorphic features and ground truth has been found economic and cheaper for proper planning and management of ground water potential zones.

The findings of the surface and ground water studies conducted through remote sensing techniques and ground truth have been discussed.

Identification and Mapping of Surface Water Bodies

Studies on the identification, location and mapping of surface water conducted in different parts of the Rajasthan desert by using remote sensing techniques in conjunction with ground turth, have revealed that water bodies of different shape and size occur under different geomorphological settings. In the middle Luni basin, small, medium and large size tanks of 1 to 2m, 2 to 3m and 3 to 10m depth largely occur in flat rocky/gravelly and buried pediments and flat interdune and older alluvial plains. These tanks due to better resolution distinctly appear on aerial photographs in dark grey to dark tone. Small tanks due to poor resolution could not be identified from Landsat imagery but the tanks of medium and large size distinctly appear in dark tone on Landsat imagery, particularly on Landsat band 7 image of wet season (Fig. 9.4). Shallow and deep and clear and turbid water bodies could be easily differentiated and mapped due to distinct tonal variations. Shallow and turbid water bodies exhibit higher infra-red reflectance and appear in white to grey tone on Landsat MSS band 7 image and in light blue tone on false colour composite. Deep and clear water bodies due to lower infra-red reflectance exhibit dark grey to dark tone on Landsat MSS band 7 image and dark blue tone on false colour composite.

In Jodhpur district of the western Rajasthan, based on the interpretation of aerial photographs and Landsat imagery, 30 major and 1554 small tanks locally known as *nadis* have been identified and mapped (Singh, 1985). These tanks largely occur in the rocky/gravelly and flat buried pediments, flat older and younger alluvial plains and flat interdune plains. The water storage capacity of major tanks varies from 0.10 to 52.84 million cubic metre (mcm) and of small tanks ranges between 0.19 to 5.08 mcm. Medium to large tanks constructed along the rivers and streams are distinctly visible on aerial photographs and Landsat imagery in dark grey to dark tone. Minor tanks due to poor resolution of Landsat MSS imagery could not be easily and accurately identified.

Landsat TM false colour composite with better resolution 30m enabled to identify and map 135 small water bodies of less than 0.9 ha and to compute their surface water and drainage basin areas and volume of water (Table 9.11). These water bodies on Landsat TM false colour composite of wet season appear in light blue to dark blue

Fig. 9.4. Landsat band 7 black and white image showing size, shape, tonal and spectral characteristics of surface water bodies

tone. The estimation of water surface area was within ±10% of the actual field estimation and was within the desired accuracy limit as recommended by White and Morain (1979). It was also observed that the buried pediments generated the highest runoff (5.5 cm) followed by the older alluvial plains (2.9 cm); rocky/gravelly pediments (1.6 cm) and interdunal plains (1.3cm). In the district, first two are the potential geomorphic sites for harvesting surface runoff in small tanks (Sharma *et al.*, 1989).

Table 9.11. Water surface and drainage basin areas of water bodies (*nadis*) under different landforms

S. No.	Landforms	No. of samples	Water surface area (m^2)	Volume of water (ham)	Drainage basin area (ha)
1	Oldalluvial plains	91	7163 (781-37422)	2.1 (0.2-14.9)	72.2 (5.0-242.5)
2	Interdunal plains	14	4626 (2500-9824)	1.8 (0.4-3.9)	137.5 (27.5 - 447.5)
3	Rocky/gravelly pediments	8	9162 (4766 - 20469)	3.7 (1.0-10.9)	231.9 (30.5 - 690.0)
4	Buried pediments	22	9436 (1719-59844)	4.2 (0.3-39.0)	76.1 (20.0-225.0)

Figures in parentheses are range values.
Source : Sharma *et al*. (1989)

In Nagaur district of the western Rajasthan, the study on the distribution, size and capacity of tanks located under different landforms were conducted by using aerial photographs and ground turth (Murthy *et al.,* (1980). The results of this study are presented in Table 9.12

Table 9.12. Distribution, size and density of tanks under different landforms

S.N.	Landforms	Area (km^2)	Number of different size of tanks			Total No. of tanks	Total Capa-city (mcm)	Density of tanks/ 100 km^2
			Large	Medium	Small			
1	Flat older alluvial plains	7778.4	327	75	167	569	38.69	7.3
2	Salder alluvial plains	114.7	7	2	1	10	1.26	6.9
3	Sandy undulating older alluvial plains	4581.1	106	69	199	374	5.15	8.2
4	Rocky/gravelly pediments	668.5	37	3	6	46	1.81	6.9
5	Flated pediments	1558.5	68	2	13	83	4.73	6.5
6	Sandy undulating buried pediments	521.8	14	2	3	19	0.82	3.7

Source : Murthy*et al*. (1980)

It is seen (Table 9.12) that the largest number of tanks with storage capacity of 38.69 mcm occur in flat older alluvial plains followed by sandy older alluvial plains having 375 tanks with 5.15 mcm storage capacity.

The Earth Resources Technology Satellite-1 (ERTS-1) re-designated as Landsat-1, Multispectral Scanner (MSS) images enabled rapid location of surface water and preliminary evaluation of surface water conditions in north-eastern South Dakota, U.S.A. (Moore *et al.,* 1974). To study the distribution, occurrence, shape and size of surface water bodies, Landsat MSS 7 band was found effective because water exhibits very low reflectance in this band. The fallow fields due to similar reflectance appeared like water bodies. To reduce the amount of non-water areas as surface water, the data should be obtained during the period of the year when the least number of fallow fields are present. Vegetated landscapes normally have higher infra-red reflectance than non-vegetated landscapes and do not appear similar to the surface water on Landsat MSS band 7 image. Surface area of the water bodies could be estimated from Landsat enlarged prints of band 7 by using compensating polar planimeter.

The Indus river valley of Pakistan experienced one of the largest floods on record. Landsat data of pre-flood and flood period made it possible to easily measure the extent of flooding, totalling about 20,000 sq. km within an area of about 400,000 sq. km south from the Punjab to the Arabian Sea (Deutsch and Ruggles, Jr. 1973). The temporal false colour composite prepared by additive projection of two band 7 scenes of September, 1972 and 1973 using red and green filters; respectively depicted the flood areas in red. The resulted temporal composite showed excellent differentiation betweeen dry areas and flood areas. In this rendition, where there was water in both the dry and flood scenes, the area was black. Where there was no water in dry scene but water during flood, the area was red. The satellite data and the rapid, economic and relatively simple optical processing techniques employed for this study make it possible to view synoptically not only the flooded but also normal water distribution, over a huge area. The remote sensing, thus helps in identification and mapping of the extent of standing flood water.

A comparative study of hydrological parameters by using Landsat MSS and Landsat TM imagery done in test area of Northwales, U.K. has revealed that Landsat TM has significant advantages over Landsat MSS in identifying and mapping water bodies like streams, lakes and tanks (France *et al.,* 1986). Lakes as small as 0.6 hetares can be identified and delineated in forested area using Landsat TM imagery due to its better resolution than Landsat MSS imagery. In the areas of open moorland, however, water bodies as small as 0.3 hectares and small streams could be identified from Landsat TM imagery.

Based on the visual interpretation of Landsat black and white imagery of band 7, it was observed that in Hirakund command area of Orissa State the water bodies less than 1 hectare cannot be located and mapped on above image (Chari *et al.* 1980). However, the water bodies larger than 2 hectares could be easily identified and mapped.

Delineation of Watersheds and Evaluation of Surface Water Potentials

Interpretation of aerial photographs and Landsat imagery helps in delineating the boundaries of watersheds of different size and shape quickly and accurately to compute their morphological characteristics and to evaluate the surface water potentials (Goose *et al.,* 1967; 1969; Singh *et al.,* 1969-71). The study on the quantitative analysis of the morphological characteristics of the Jawai and the Banas watersheds located in different parts of the Rajasthan revealed that the watersheds occurring on the phyllite, schist and rhyolite pediments have the greater values of bifurcation ratio, drainage density, number of streams, stream frequency of first order segment over total basin area. The higher values of morphological characteristics indicated that these watersheds due to circular shape, shorter stream lengths, steep gradient and closely spaced and integrated drainage network will have better surface water potentials for development. The watersheds developed on the granite pediments and sandy older alluvial plains will generate poor surface runoff due to longer stream lengths, elongated shape, gentler gradient, widely spaced drainage channels and poor drainage density.

Morphohydrological investigations of watersheds in sandstone and limestone region of Jodhpur district revealed that aerial photo-interpretation can be successfully used to measure the morphological characteristics of watersheds. These morphological characteristics in turn influence the hydrological characteristics of the watersheds (Singh and Sharma, 1979). The surface water potentials worked out from these watersheds varies from 6.59 to 82.04 ha metres. It has been qualitatively evaluated that the surface runoff increases with the increase of total basin area, total stream length and drainage density. The watersheds with the largest area will have the highest runoff potentials. The surface water from potential watersheds could be utilised for storage and ground water recharge at suitable geomorphic sites.

The runoff potentials of the Kolayat and Nal watersheds situated on eroded rhyolite pediment and eroded older alluvial plain, respectively in Bikaner district of western Rajasthan have been worked out by Yadav *et al.* (1976). From the results of this study, it has been inferred that Kolayat watershed has runoff potentials of 1600 ha metres in deficit year and 8820 ha metres in flood years. The runoff potentials of Nal watershed in normal year are 1000-1300 ha metres and in flood year are 11500 ha metres. The runoff potentials from these watersheds could be utilised for storage, irrigation and drinking.

Satellite remote sensing is useful for surface water planning as it is a reliable to map rivers, streams and other surface water bodies and to extract values of the parameters used in hydrological models (Drayton, 1986). The visual interpretation of satellite imagery can provide drainage network and reliable data for stream flow models at low cost. The single band (near infra- red) and false colour composites are suitable for hydrological work like delineating the boundaries of the watersheds and to estimate their catchment area. Although the details of the drainage channels on satellite imagery at many places are wrong but the major boundaries between drainage units can be plotted with care and accuracy. Hence, it is concluded that the use of this cheap and simple technique is useful and appropriate to the need of the third world.

In Wyoming, U.S.A., under a semi-arid climate, the use of aerial photographs has been made to compute the geomorphic characteristics of the watersheds and to correlate them with the hydrologic characteristics (Hadley and Schumm, 1961). In this region, a good correlation between drainage density and mean annual runoff was established. This correlation showed that drainage density could be used to predict annual runoff from drainage basins in eastern Wyoming.

Identification, Delineation and Mapping of Ground Water Zones (Aquifers)

The studies conducted on identification, delineation and mapping of ground water zones (aquifers) in different parts of the Rajasthan desert using aerial and satellite remote sensing techniques through random field checks have revealed that the development of potential aquifers is governed by the geomorphic features of the region which are in turn controlled by lithology, rock structure and drainage conditions (Chatterji *et al.* 1979; Singh, 1976-77, 1981).

In the central Luni basin, 12 aquifers viz., blown sand, younger alluvial, older alluvial, Vindhyan sandstone, Malani volcanic, Siwana granite, Jalor granite, Adamellite, quartzite, Aravalli slate and phyllite and granite gneiss have been indentified, delineated and mapped (Fig. 9.5). Geomorphic factors such as drainage patterns, slope, nature of sediments, abandoned drainage channels, levees, rock patterns, lineaments and altitude of landforms which control the occurrence and development of these aquifers could be easily interpreted from aerial photographs and Landsat imagery. The first three types of aquifers are shallow and their water bearing formations are sands, older alluvium and younger alluvium. Due to thick impervious nature of the water bearing formations, these aquifers have good ground water potentials and water occurs at 6 to 75 m depth. Remaining nine types of aquifers are deep to very deep and the water bearing formations are weathered and solid rocks of different types. Sandstone aquifers occur under rocky/gravelly and buried pediments, older alluvium and sand dunes and have good ground water potentials due to well developed lineaments, dendritic and parallel drainage patterns and weathered rocky strata. The depth of ground water due to varying depth of water bearing strata ranges from 20 to 46 m below ground level. The aquifers developed in

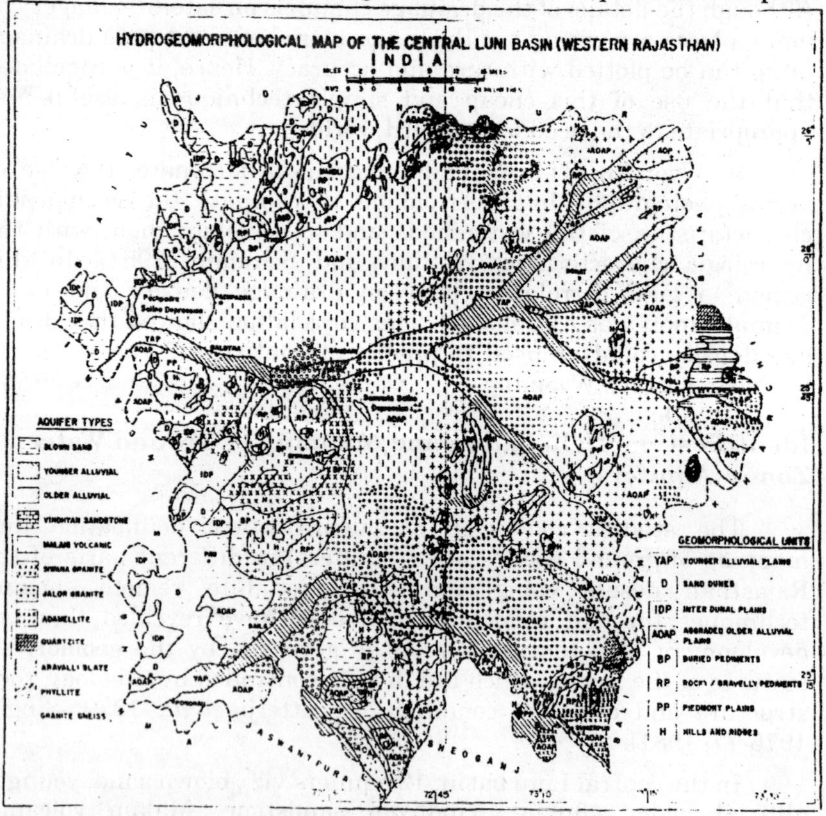

Fig. 9.5. Hydrogeomorphological map of the central Luni basin (western Rajasthan), India

the igneous and metamorphic rocks occur under different landforms like/rocky gravelly and buried pediments, and older and younger alluvial plains. These aquifers except granite are not potential. The static water level of such aquifers ranges from 40 to 100 m below ground level.

The courses of prior buried drainage channels of the Luni and the Ghaggar drainage systems have been reconstructed and mapped from the aerial photographs and Landsat imagery with the help of scar marks and tonal variations due to the better stand of vegetation and moisture conditions than surrounding areas (Singh, 1976-77; Ghose *et al.,* 1979). The rain water through the buried courses of the prior drainage channels filled with impervious sediments like sand, gravel and alluvium flows subterraneously resulting into the development of potential aquifers. The disorganised drainage channels regulate the subterraneous flow of ground water and water here occurs at 4 to 15 m depth below ground level. The distribution of ground water along the present and prior drainage channels in the Central Luni basin has been mapped (Fig. 9.6).

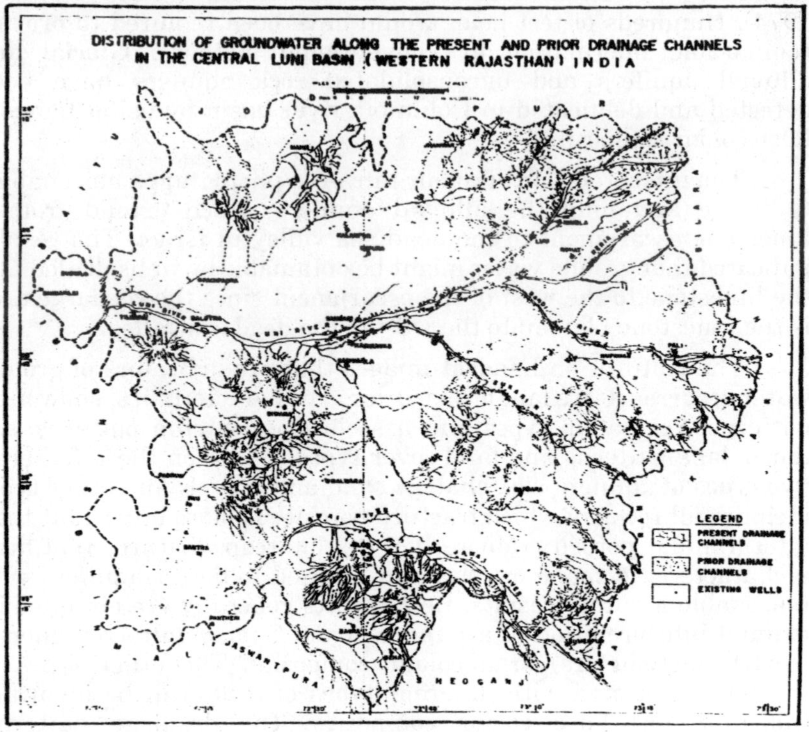

Fig. 9.6. Distribution of groundwater along the present and prior drainage channels in the central Luni basin (western Rajasthan), India

Remote sensing investigations of the greater Chaliyar basin of the Kerala State revealed that the predominant lineament direction in this basin is NE-SW and NW-SE and hence the rectangular open wells in this area should have their lengths prependicular to these directions (Basak *et al.*, 1986). The density of lineaments in different sub-basins indicate that the density is more in the Kallai river basin, followed by Kadalundi, Korapuzha and Chaliyar. From this, it can be stated that the potentiality of ground water will be comparatively more in Kallai river sub-basin followed by Kadalundi sub-basin. Based on the presence of lineaments, distribution of valley areas (Elas) and other geomorphological features, favourable locations for ground water exploration in the Chaliyar basin have been suggested. The suggested potential sites should be ascertained by geophysical studies before drilling activities are undertaken.

Application of remote sensing techniques were used to determine the distribution of ground water in a discontinuous glacial-drift mantle over the very rugged surface of the iron-bearing formation in Escanaba river basin, Michigan, U.S.A. (Deutsch, 1974). Hundreds of test holes would have been required to provide comparable information by conventional methods. Glacial and alluvial aquifers and unconsolidated rock aquifers have been detected and delineated in Columbia river basin by using Landsat false colour composites.

The interpretation of colour infra-red photographs has enabled to locate and map the alluvial aquifers which extend from a limestone escarpment to the dead sea valley in Israel. The results indicated that ground water might be obtainable by wells drilled into the limestone to the west of the escarpment since the recharge area is the limestone plateau to the west of the dead sea valley.

The features on Landsat images that are indicators of ground water occurrence are landforms, drainage characteristics, snow melt patterns, vegetation types and associations, outcrop patterns, soil tones, lake patterns and land cover types. Based on these features, two types of aquifers viz., shallow sand and gravels and other types includes all rocks in which fracture porosity is more important than intergranular porosity (Moore, 1977). Landscape features that have a characteristic shape or form can be used to detect aquifers with reasonable accurate results. Best results generally are obtained by manual interpretation of a Landsat band 7 (near infra-red energy, 0.8-1.1 micrometers) or a colour composite. The other types of aquifers associated with different types of rocks can be identified from Landsat false colour composite. The drainage patterns, vegetation patterns, folds and lineaments control the development and occurrence of such aquifers which could be easily identified and mapped from Landsat false colour composites and band 7 image.

Side Looking Airborne Radar (SLAR) was used to locate potential sources of fracture controlled ground water in pre-Cambrian crystalline rocks of north-central Nigeria (Gelmett and Gardner, 1979). Lineaments of three kilometres and of greater length were interpreted from the imagery, statistically analysed and calssified to determine potential well-site location at the intersections of, or along, preferred tensional sets. The above radar with its synoptic format and the proper use of depression angles is superior to other senser systems for mapping lineaments and other structural features. The potential offered by SLAR for ground water exploration and development in the drier regions of the world has been clearly demonstrated.

Detailed geomorphological mapping of fracture traces and lineaments, basement fractures, drainage patterns, palaeo-drainage, folds, graded channels and alluvial fan through aerial and space photographs enabled to locate and delineate potential ground water zones in eastern parts of the Rajasthan. Results of the study revealed that the aerial and satellite images when used in conjunction with geophysical and geological mapping help in area selection for ground water exploration, pin pointing of well drilling sites and acquisition of certain remote sensing based parameters of use in hydrologic equations for assessment of ground water resource potential.

Digital Analysis of Landsat Data for Water Resources Evaluation

Digitally edge-enhanced IRS-LISS II and Landsat TM FCC proved very useful for identification and mapping of water bodies like village tanks (*nadis*), reservoirs and rivers in the south-eastern part of Jodhpur District, western Rajasthan. Based on the tonal variations of the water bodies, their depth could be qualitatively estimated and temporal changes in surface area could also be detected. The density of village tanks (*nadis*) is the highest in the rocky/gravelly pediments, followed by the flat older alluvial plains.

Lineaments and buried courses of the prior drainage channels which are considered as the significant indicators of ground water potential zones could be identified and mapped from digitally edge-enhanced IRS-LISS II and Landsat TM False Colour Composites (FCCs). Three major ground water zones viz., younger alluvium, older alluvium and granite could be identified and delineated for systematic ground water exploration (Kar *et al.*, 1989).

In Karnataka State, two case studies conducted in Lidda-ghatta- Vijayapura area and Shimsha basin area have shown that

Landsat imageries are extremely useful in indentification of water bearing zones in crystalline rocks. Landsat false colour composites, linear stretch and band ratio images were studied for increasing the interpretability facets, fractures, fissures and drainage patterns to locate potential ground water zones (Reddy and Gaikwad, 1986). It has been observed that linear stretched images of Landsat band 5 and 7 enhanced the linear features, lithology and drainage. In the enhanced FCC (band 4, 5 and 7), drainage, linear features and lithology are very clearly visible.

From the interpretation of digitally enhanced Landsat images, the areas feasible for ground water exploration and exploitation have been identified in the above two areas.

INTEGRATION, ASSESSMENT AND MANAGEMENT OF NATURAL RESOURCES

The physical and biological natural resources of the Jalor district have been integrated in the form of Major Land Resources Units (MLRUs) (Fig. 9.7). Each MLRU is unique with respect to its natural resources and, therefore, has similar resource potential and management needs (Singh *et al.,* 1995). The location, salient characteristics, assessment and general recommendations for the sustainable development of each MLRU are discussed below.

MLRU 1. Hills (249 km^2)

Characteristics : The rhyolite and granite hills covered with skeletal soils occur near Bhadrajan, Malgarh, Jalor, Roza, Israna, Jaswantpura and Dorra villages and have been classified in class VIII land. Dominant vegetation communities on the hills are *Anogeissus pendula* - *Acacia senegal, Euphorbia caducifolia* - *Commiphora wightii* and *Aristida oropetium*. The dominant landuse is forest under various degree of degradation.

Assessment : The hills are subjected to severe water erosion and have high run-off potential with run-off coefficients of 0.6 to 0.8 with practically no groundwater resources. This run-off recharges the ground water in the vicinity. The granite hills are source of famous Jalor granite. Quarrying on large scale has degraded flanks of the hills. At the western flanks of these hills, thick aeolian sands in the form of sand dunes and hummocks have been deposited which are dissected by the water in the form of deep gullies.

Recommendations : At the granite quarry sites, the suitable plant species may be raised by adopting water harvesting technique. The gullied areas at the flanks of hills may be rehabilitated by

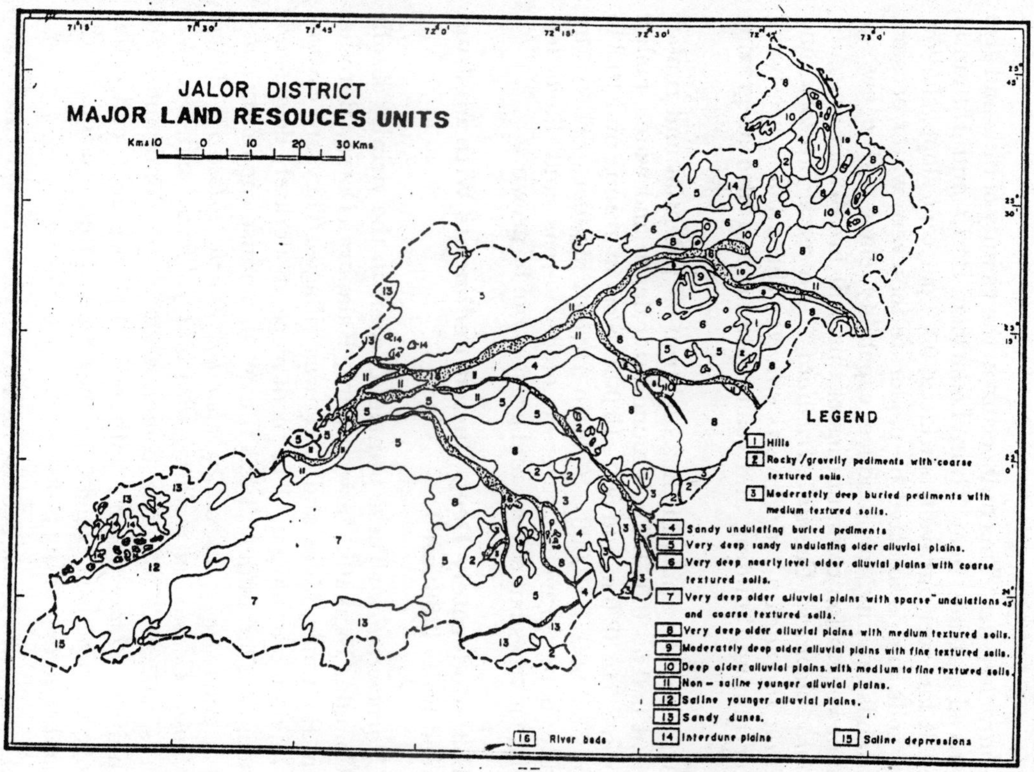

Fig. 9.7. Major land resources unit (Jalor district)

plantation of suitable species and proper soil and water conservation measures. The hill slopes and the tops need plantation of suitable tree species.

MLRU 2. Rocky/Gravelly Pediments With Coarse Textured Soils (110 km^2)

Characteristics : This unit occurs in the vicinity of hills and also in scattered patches near Thur, Ledarmer, Borta and Hatamtai villages. It has rocky/gravelly surfaces, 0 - 2 per cent slope, shallow skeletal soils and is affected by moderate to severe water erosion. These lands qualify for landuse capability class VII c ew sh. Dominant vegetation communities of this unit are *Acacia senegal - Salvadora oleoides, Cassia auriculata - Prosopis juliflora* and *Oropetium thomaeum - Eragrostis ciliaris*. These lands are being used for mining, building and road material and also as grazing lands.

Assessment : This unit has high run-off potential and run-off coefficient is 0.5 - 0.060. The groundwater depth and water quality vary form 8.5 to 32.0 m and 800 - 15,000 S/cm in granite and 10.3 to 29.0 m and 900 - 12,000 S/cm in rhyolite aquifers. These are generally bare, gravelly, severely eroded areas and have low potential for trees but shrubs and grasses can be grown.

MLRU 3. Moderately Deep Buried Pediments With Medium Textured Soils (398 km^2)

Characteristics : These pediments occur in the vicinity of hills near Goindla, Bankalli, Jaswantpura, Kalapura, Dorra, Baretha, Silasan, Rampura, Mera and Lackawas villages. They are covered with alluvium and have 0 - 1 per cent slope, moderately deep sandy loam soils and fall under VI & VII c ew sh landuse capability class. *Prosopis cineraria - Salvadora oleoides, Acacia nilotica - Acacia leucophloea, Cassia auriculata - Prosopis juliflora* and *Dactyloctenium sindicum - Eleusine compressa* are the dominant plant communities. Monocropping with 80 - 100 per cent cultivation intensity is practised. Slight to moderate water erosion is the dominant land degrading process.

Assessment : These pediments have shallow to moderately deep gravelly soils often subjected to moderate to severe water erosion. This unit has a moderate run-off potential with run-off coefficient of 0.4 - 0.5. The groundwater is situated at 20 - 40 m depth and water quality ranges from 2000 - 4000 S/cm.

Recommendations : Contour bunding and other suitable soil and water conservation measures as per the requirements of site should be adopted. Silvipastoral and silvihortipastoral is the ideal land use for this unit.

MLRU 4. Sandy Undulating Buried Pediments (252 km^2)

Characteristics : This unit occurs in Bala, Korana, Kishangarh, Pavali, Panseri, Daspan, Pantheri and Sendara villages. Moderately undulating topography and very deep fine sandy soils are the characteristic features of this unit which qualify for landuse capability class IVc ea s. *Prosopis cineraria - Tecomella undulata - Maytenus emarginatus* form the dominant plant community. Monocropping with 60 - 80 per cent cultivation intensity is done. Moderate to severe wind erosion/deposition, and water erosion are dominant land degradation processes.

Assessment : This unit is vulnerable to wind and water erosion. Due to undulating surface and hummockiness, these lands are not well suited for cultivation. There is moderate run-off potential in pockets with a run-off coefficient of 0.2 - 0.5. The ground water depth and quality ranges from 10 - 30 m and 200 - 4000 S/cm, respectively. The soils are deep having moderate water holding capacity and fertility status.

Recommendations : These lands need protection against severe wind and water erosion and use of heavy machines on them should be avoided. The shrubs and grasses should be grown on the low low dunes and hummocks. In flat areas, rainfed cropping in combination with grass strip may be taken up. Silvipastures and hortipastures are the ideal land uses for this unit.

MLRU 5. Very Deep Sandy Undulating Older Alluvial Plains (1863 km^2)

Characteristics : This unit occurs near Raithal, Odwara, Sarana, Saila, Sirana, Mank, Daspan, Jetu, Jeran, Jetu, Karara, Dantiwa and Punasa villages. Undulations due to hummocky surface, very deep and fine sandy soils of IVc ea s landuse capability class are the characteristic features of this unit. *Acacia nilotica-Prosopis cineraria-Acacia jacquemontii, Dactyloctenicum sindicum, Eleusine compriessa* form the dominant vegetation community. The monocropping with 60 - 80 and 30 - 60 per cent cultivation intensity is practised. The area is degraded due to moderate to severe wind erosion/deposition.

Assessment : This MLRU has good potential for supporting various type of trees and shrubs. Because of sandy soils and hummocky surface these lands don mt permit use of heavy machinery. This unit has poor run-off potential with a run-off coefficient of 0.10 - 0.15. The ground-water which occurs at 10 - 40 m depth is of 2000 - 4000 S/cm electrical conductivity.

Recommendations : Silvipasture in combination with rainfed cropping is an ideal landuse for this unit. On the hummocks and along field boundaries and also against the direction of wind, tree species like *Prosopis cineraria, Salvadora oleoides, Acacia tortilis* and shrubs like *Capparis decidua* and *Clerodendrum phlomoides* should be widely planted in the flat area of this unit. Rainfed crops like *bajra, moong* and *moth* may be taken up. The wind erosion control measures should form the integral part of management.

MLRU 6. Very Deep Nearly Level Older Alluvial Plains With Coarse Textured Soils (682 km^2)

Characteristics : This unit occurs in Debawas, Sarana, Dudia, Beria Devaki, Mera Uparla and Dhowala villages. Nearly level surfaces, loamy sand soil and IIIc ea and IVc ea land capability classes are the dominant features of this unit. The dominant plant communities are *Salvadora oleoides-Prosopis cineraria, Capparis decidua- Prosopia juliflora* and *Dactyloctenium sindicum-Eleusine compressa*. Monocropping with 60 - 80 per cent and 80 - 100 per cent cultivation intensity is practiced in this unit. The region is degraded due to moderate wind and water erosion.

Assessment : The soils are very deep, loamy sand and at places irrigated with moderately brackish to saline water. These soils are well suited for irrigated and rainfed cropping. The region has a poor run-off potential with run-off coefficient of 0.10-0.20. The depth to groundwater is 10-20 m with 4000-8000 S/cm electrical conductivity.

Recommendations : Use of fertilizers and FYM, both under rainfed and irrigated conditions, are recommended. Wherever irrigation with high RSC water is being followed, gypsum should be applied as per requirement to maintain productivity of the land. The *bajra, moong* and *til* under rainfed condition and cumin and *isabgol* under brackish water irrigation can be grown.

MLRU7.VeryDeepOlderAlluvialPlainsWithSparse Undulations and Coarse Textured Soils (1896 km^2)

Characteristics : This unit mainly occurs near Karara, Aranay, Amali, Kachhela, Vank, Pahadpura, Sidhesar and Sarnau villages.

The surfaces of these plains are flat interspersed with fenceline hummocks, very deep fine sandy soils having landuse capability classes of IIIc ea s and IV c ea s. The dominant plant communities of this unit are *Prosopis cineraria-Salvadora persica, Prosopis juliflora-Crotalaria burhia* and *Eragrostis ciliaris-Cenchrus setigerus*. Monocropping with 20 to 30, 30 to 60 and 60 to 80 per cent cultivation intensity, respectively is practiced. Part of this unit is also permanent pasture/*oran*. Moderate wind erosion/deposition is the major hazard of this resource unit.

Assessment : The area of this unit is generally flat having scattered dunes and low hummocks. This unit has a poor run-off indicated by run-off coefficient of 0.1 - 0.2. The groundwater at 10 - 20 m depth has 4000 - 8000 S/cm electrical conductivity. Because of poor quality and deep ground water there is not much scope for irrigation. The soils are prone to wind erosion and requires careful management.

Recommendations : This unit is suitable for rainfed cropping with improved soil and moisture conservation measures. The soils are well suited for rainfed cropping, afforestation and grassland development, particularly in areas where moderate to severe wind erosion has degraded the land. Application of fertilizers and farm yard manure with improved varieties of crops under rainfed and irrigated condition will improve the crop yield to a great extent. Hortipastoral, agrihorticulture and silvipastoral practices should also be adopted.

MLRU 8. Very Deep Alluvial Plains With Medium Textured Soils (2281 km^2)

Characteristics : This unit occurs near Kamba, Ajitpura, Ukharra, Mulewa, Bijali, Barwa, Bala, Sandan, Devara, Bibalsar, Mandoli, Ramsin, Chandana, Bara, Dhansa, Ruchiyar, Bhimpura; Kushalpura, Jetu, Mundawas and Jeran villages and is characterized by flat surfaces and very deep loamy sand to sandy loam soils of landuse capability classes IIIc ea s and IVc sa. *Acacia nilotica-Salvodara oleoides, Ziziphus nummularia- Capparis decidua* and *Dactyloctenium sindicum-Eleusine compressa* are the dominant plant communities. The dominant landuses are double cropping and monocropping with 60 - 80 per cent cultivation intensity. Some areas of this unit are affected by moderate water and wind erosion/deposition.

Assessment : These are very potential areas for irrigated and rainfed cropping. The region has a poor to moderate run-off potential as indicated by run-off coefficient of 0.1 - 0.3. The ground water is encountered at 10 - 20 m and 20 - 40 m depth and has a quality ranging from 2000 - 4000 S/cm and 6000 - 8000 S/cm, respectively. The soils have good available moisture capacity. These lands are moderately prone to wind and water erosion. At certain places brackish ground water is also available for irrigation.

Recommendations : Crops in *kharif* season viz. *bajra, guar* and *moong* can be raised by adopting *in situ* water harvesting techniques. Under irrigated situations, wheat, *rayda (Brassica* sp.) cumin can be grown. Fertilizer and FYM application and use of plant protection measures will greatly enhance the crop yields. Fruits trees like *ber*, guava, *sitaphal* (custard), *gonda* and *karonda* should also be planted wherever water for irrigation is available. There is a good scope of raising the yields of cumin and *isabgol* by adopting improved techniques.

MLRU 9. Moderately Deep Older Alluvial Plains With Fine Textured Soils (48 km^2)

Characteristics : This unit mainly occurs near village Bhanwarani and occupies vast area. Almost level surfaces, loam to clay loam soils and landuse capability class IIIc are the salient features of this unit. This vast plain is devoid of trees. The dominant vegetation includes *Indigofera oblongifolia, Capparis decidua* and *Cenchrus ciliaris*. In general, the monocropping with 80 to 100 per cent cultivation intensity, is practiced.

Assessment : In this unit, the lands are level with fine textured soils receiving run-off water during rains. The area has moderate run-off potential and run-off coefficient of 0.2 - 0.4. The groundwater at 10 - 30 m depth has 2000 S/cm electrical conductivity. These soils have high available water holding capacity and good agricultural potential.

Recommendations : These plains don mt have any physical limitation and under conserved moisture, wheat, *rayada* and gram can be grown in *rabi* season. More run-off should be conserved and the acreage under crops like *rayada* and wheat grown on the conserved moisture may be increased. The crop yields can be improved by application of FYM and light dose of fertilizers. The area remains flooded during rains and, therefore, trees which could survive under waterlogging conditions may be grown.

MLRU 10. Deep Older Alluvial Plains With Medium to Fine Textured Soils (463 km^2)

Characteristics : This MLRU extends in Ahor, Bithure to Haryali, Balotan, Bithura, Goliya, Takhatgarh, Norwa, Devgarh,

Sankhwali, Guda Rama and Dudia villages. Nearly level surfaces, loam to clay loam deep soils belong to landuse capability class IIIc and IIIc sa. *Acacia nilotica-Salvadora oleoides, Indigofera oblongifolia-Tamarix aphylla-Prosopis juliflora* and *Cynodon dactylon-Eleusine compressa* are the dominant plant communities.

Assessment : This unit due to the deep fine textured soils, absence of the wind erosion/deposition and water erosion hazards and ground water availability has better agricultural potentials. These areas have moderate run-off potential and run-off coefficient of 0.1 - 0.3. The ground-water is saline EC (4000 - 8000 S/cm) and occurs at 10 - 30 m depth. Double cropping and monocropping with 80 - 100 per cent cultivation intensity are commonly practised here. Because of good ground water potential, a number of wells occur in this unit and irrigated agriculture is practiced and good yields of wheat, cumin, *rayada*, cotton and vegetables are obtained. However, at some places water is sodic which adversely affects soil properties and crop yields.

Recommendations : Under conserved moisture, wheat, *rayada* and gram can be taken. Wherever water for irrigation is available, the FYM and fertilizer should be applied and crops like cotton, wheat, *rayada*, cumin and *isabgol* should be grown. In sodic water irrigated fields, gypsum should be applied to counter is negative effect and crops wheat and barley which can tolerate adverse effects of RSC waters should be grown.

MLRU 11. Non-Saline Younger Alluvial Plains (833 km^2)

Characteristics : This MLRU occurs along the rivers banks near Haryali, Ahor, Otwal, Turan, Unri, Deta, Surana, Dadal, and Dhumbadiya villages. Almost level surfaces with less than 1 per cent slope, deep to very deep, fine sand, fine sandy loam and silty loam soils belong to landuse capability class IIIc. *Salvadora oleoides-Prosopis cineraria, Prosopis juliflora- Cassia auriculata* and *Cynadon dactylon-Desmostachya bipinnata* are the dominant plant communities. Rainfed and irrigated farming are practised here. Moderate water erosion as well as inundation in pockets are the major degradation processes of this unit.

Assessment : The sediments of this unit are deep, variable in texture and non-saline alkali. These areas are frequently inundated by the rivers during rainy season. The region has moderate run - off potential with run - off coefficient of 0.1 - 0.3. Depth to the ground water ranges from 410 - 420 m and its quality due to 1000 - 2000 S/cm EC is good. These areas are intensively cultivated for vegetables, fruit and flower trees and irrigated crops like wheat, *rayada*, cumin and *isabgol*.

Recommendations : Cash crops like cotton, chillies, cumin, mustard by adopting improved seeds, fertilizers and plant protection measures should be taken up. Because of availability of good quality ground water at shallow depth, this unit offers potential for vegetable and flower cultivation and also raising orchards of guava, *ber*, pomegranate, *gonda, karonda* and *sitaphal* (custard). Balanced fertilization including application of potassium should from a part of regular practice.

MLRU 12. Saline Younger Alluvial Plains (346 km^2)

Characteristics : This unit mainly extends in Chitalwana to Hotigaon, Meda, Bhawatra and Khejadiyali villages and has flat surface and deep soils varying in texture from fine sand to find sandy loam and silty loam. These lands have high salinity and impeded drainage and qualify for landuse capability class VIIc sa. *Salvadora persica-Salvadora oleoides, Prosopis juliflora- Capparis decidua* and *Cynodon dactylon-Desmostachya bipinnata* are the dominant plant communities of this unit.

Assessment : This unit suffers with several constraints including relatively low rainfall, frequent salinity/sodicity in soil and poor quality of ground water, stratified soils and inundation in rainy season. The region has moderate run-off potential and 0.1 - 0.3 run-off coefficient. The groundwater is saline (8000 S/cm) and occurs at less than 10 m depth. The vegetation cover is also sparse and mainly consists of *Prosopis juliflora*. Generally the area has low potentiality for agriculture.

Recommendations : During good rainfall year due to inundation, salinity is reduced in some areas where crops like wheat, gram and *rayada* may be grown on the conserved moisture. However crop yields under these situations are only moderate and hence these plains are better suited for the development of silvipastures. Based on the salinity level of the soils, the salt tolerant shrub, trees or grasses may be grown.

MLRU 13. Sand Dunes (574 km^2)

Characteristics : This unit occurs near Paladar, Nenol, Panchala, Tentop, Kuda, and Tantol villages. Stabilised sand dunes with 10 to 40 m height and very deep fine sandy soils through out the profile qualify for landuse capability class VIIc eas. *Tecomella undulata-Salvadora oleoides-Prosopis cineraria,* and *Aerva pseudotomentosa-Leptadenia pyrotechnica-Panicum turgidum* are the dominant plants communities. These dunes are put under cultivation during favourable rainfall year and monocropping with 20 - 30 and 30 - 60 per cent intensity of cultivation is practised.

Assessment : Most of the dunes are stable and fairly vegetated. At few places, the obstacle dunes have been dissected and gullies have been created. The dunes in the Sanchor area are generally devoid of natural vegetation. These are capable of supporting good vegetative cover. Moderate wind erosion/deposition is the major soil degradation process, resulting in the reactivation of the dune crests and flanks.

Recommendations : Cultivation on sand dunes should be restricted only upto lower flanks and wind ward sides of the dunes. The mobile dunes should be stabilized by the technique developed by the CAZRI. The crests and upper flanks of the dunes should be developed into silvipastures using suitable tree, shrub and grass species. The reactivated parts of the stabilised dunes should be revegetated by providing barbed wire fencing.

MLRU 14. Interdune Plains (98 km^2)

Characteristics : This unit occurs among the clusters of the stabilised sand dunes and near those villages where various type of sand dunes have been encountered. It is flat and undulating resource unit having irregular slope varying from less than 1 to 4°. The soils of these plains are very deep fine sand to loamy fine sand. The dominant type of vegetation is *Leptadenia pyrotechnica* and *Panicum turgidum*. Cropping intensity of cultivation varies from 30 - 60 and 20 - 30 per cent. But in Raniwara area, the interdune plains have good quality water at shallow depth. Moderate wind erosion is the major process of land degradation.

Assessment : The flat interdune plains due to flat topography and very limited physical hazards are fit for rainfed and irrigated agriculture. Whereas the undulating interdune plains due to various constraints like undulating topography, irregular slope, poor water holding capacity and fertility status are not suitable for cultivation but have a good potential for the development of silvipastures, using proper soil and water conservation measures. The surface run-off coefficient is 0.2 - 0.3. The ground water is 10 - 20m deep and saline (EC 6000 - 8000 S/cm).

Recommendations : In the flat interdune plains, the rainfed agriculture using improved varieties of *bajra, mung, moth* and *guar* and proper doses of fertilizers is recommended. These plains in Raniwara area can be cultivated for irrigated crops like wheat, mustard, cumin and *isabgol* but the cropping should be at low intensity. Application of FYM and fertilizers should be a regular practice. Sandy undulating interdune plains due to their physical limitations should be avoided from cultivation. These plains could be developed into silvipastures and hortipastures, using suitable fruit and non- fruit trees and grasses for meeting the demand of fuel, fodder and fruits.

MLRU 15. Saline Depressions (148 km^2)

Characteristics : This unit occurs near Mohiwara, Nilkantha, Arwa, Agdawa, Alasan, Reothara, Chawrda, Nimbla, Bhawatara and Bhataki villages. Due to almost level saline surfaces with south-west to north alignment and deep saline alkali sandy loam to clay loam soils, this resource unit qualifies for the landuse capability class VIIc, sa. *Prosopis juliflora- Suaeda fruitcosa* and *Sporobolus helvolus-Aleuropus lagopoides* are the dominant vegetation communities. The landuse is saline waste with scrub in fringe areas.

Assessment : This unit has major constraints like hard crust, impeded drainage, high salinity and calcareousness and hence permit the growth of only salt tolerant species. The groundwater due to EC values more than 800 S/cm is saline and it occurs at less than 10m depth. *Prosopis juliflora* does well and is growing in vast area of this unit. These lands can support salt tolerant grasses and shrubs.

Recommendations : The areas of this resource unit not fit for salt extraction, could be developed into pastures by growing salt tolerant plant species using proper rainwater harvesting and soil and water conservation techniques. Whereas the areas rich in salt contents should be continued for salt extraction.

MLRU 16. River Beds (399 km^2)

Characteristics : This unit occurs near Bankli, Harsi, Saila, Siyana, Thur and Badwi villages. It is characterised by ephemeral flow, coarse gravelly sands, sand bars of different dimensions presence of barchan dunes and gradient varying from 1 : 400 - 1:700. The width of beds ranges from less than 500 m to more than 5500 m, average width is 1968 m and their depth varies from 0.5 to 3 m. Due to meandering, extreme western parts are having wider widths; and base flow is absent. They remain dry for 90 per cent of the year; and support *Prosopis juliflora* and *Tamarix auriculata.*

Assessment : River beds are in a severe degraded state, bank erosion is prominent, limited potential for bed cultivation due to very coarse sandy nature of the sediment. Sands of the river beds are generally used for the construction of residental, government and private buildings.

Recommendations : In the headwater region, these river beds may be used for subterranean water extraction and subsurface barrier construction for artificial recharge of groundwater. The lateral shifting of the river beds to check bank erosion and further degradation of adjoining agricultural lands should be checked by constructing spur at 45° angle.

SUGGESTED LANDUSE PLANNING

Morphopedological characteristics, biophysical potentials and inherent vulnerability of the landforms identified in Jalor district will form a sound basis for their rational landuse planning, using suitable soil and water conservation measures (Fig. 9.8). The large areas of the fluvial landforms viz; moderately deep, deep and very deep flat older alluvial plains, flat interdune plains, flat buried pediments and younger alluvial plains are being used under rainfed farming. The production and productivity of these landforms ae depleting due to the unscientific landuse planning and increasing degradational activities. The sustainable development of these landforms need suitable measures like contour bunding, contour furrows, shelterbelts to reduce their degradation, appropriate landuse, suitable crop rotation, scientific methods of cultivation, proper maintenance of fertility status and harvesting of surface run-off in farm ponds for supplementary irrigation.

The improved crop varieties of *bajra* (HHB-60, HHB-67, WCC-75), *cowpea* (K-11), *moong* (K-851, S-8) and *guar* (Maruguar, Durgapura safed) and *moth* (Marumoth, RMO 48) with use of 30 - 50 kg/N/ha and 30 - 40 kg/ha P_2O_5 will upgrade and stabilise the productivity of these landforms. The dry farming practices like contour or field bunding to minimise run-off losses, bentonite clay as subsurface barrier to reduce moisture losses, interplot water harvesting for improved moisture regime and sowing with seed-cum fertilizer drill in a paired or uniform row system with proper spacing have been suggested for upgrading the yield of the rainfed crops. The intercropping of pearl millet with grain legumes (green gram, clusterbean and mothbean) has been recognised a viable system for sustainable development of different landforms.

The saline flat older alluvial plains unfit for cultivation could be best developed into silvipastures with salt tolerant species like *Tamarix articulata, Prosopis tamaruga, Atriplex nummularia, Dichanthium annulatum* and *Sporobolus helvolus*.

Aeolian landforms viz; sandy undulating buried pediments, sandy undulating older alluvial and interdune plains due to inherent vulnerability of the sandy soils, irregular slope, poor fertility status and moisture regime should not be used for cultivation. In order to stabilise the production and productivity of these landforms, the hortipastoral and agrihorticulture systems may be introduced with proper soil and water conservation measures like wind breaks, shelterbelts, contour furrows and field bundings. The sand dunes

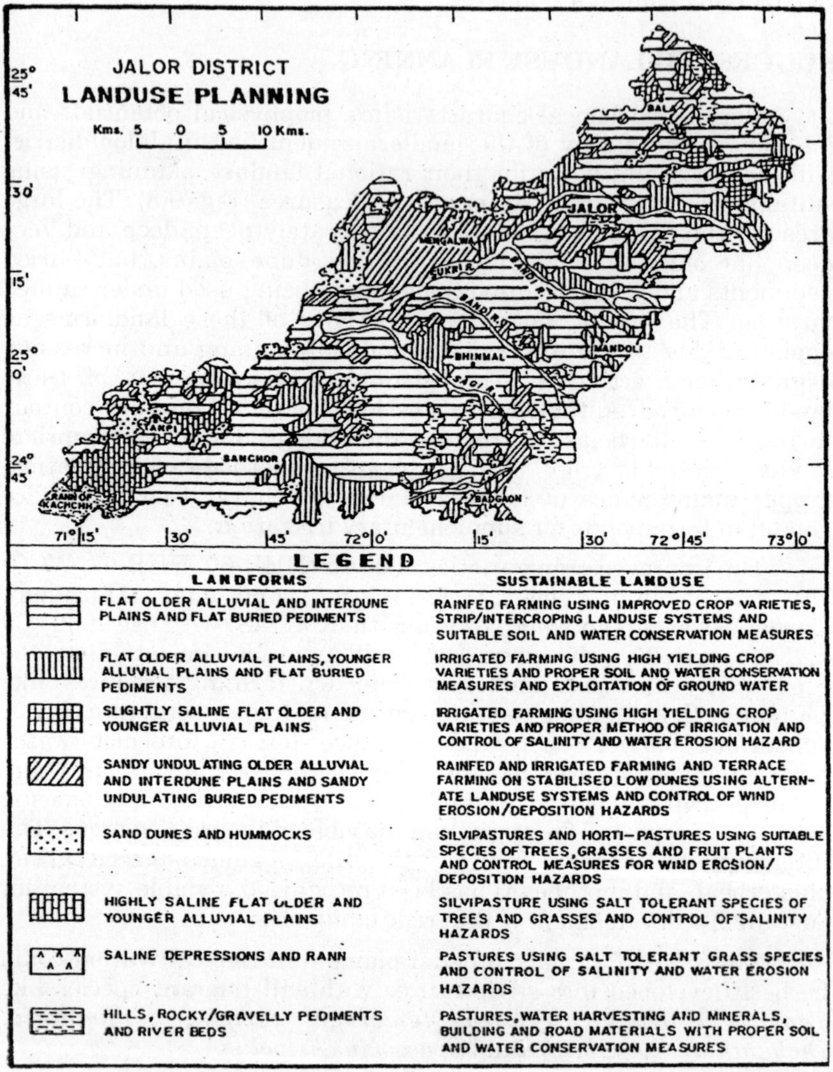

Fig. 9.8. Landuse planning (Jalor district)

and undulating areas of the above landforms are best suited for the development of silvipastures instead of cultivation with suitable plant species like *Lasiurus sindicus, Cenchrus ciliaris, Prosopis cineraria, Acacia senegal, Calligonum polygonoides* and *Ziziphus nummularia.*

The production and productivity of fluvial and aeolian landforms where irrigated farming is practised could be increased by using high yielding crop varieties, proper method of cultivation and suitable soil and water conservation measures like wind breaks, wind strip cropping, contour bunding and field bunding. More acreage of these landforms could be brought under irrigated farming by exploiting ground water resources in the under exploited zones located in Ahor and Jalor tehsils. In Sanchor tehsil, the stabilised low dunes are being used under terrace farming for growing cash crops and more areas could be used for above type of farming through judicious use of ground water and proper soil and water conservation measures.

The hills and rocky/gravelly pediments due to severe to very severe physical limitations are not fit for cultivation. They should, therefore, be developed into pastures, silvipastures, desert catchments and mining by using proper soil and water conservation measures like contour bunding, gully plugging and grass cover on mine spoils to check wind erosion.

The small and large saline depressions, not fit for cultivation, could be reclaimed and developed into pastures by using halophytic plant species such as *Sporobolus marginatus, Cressa cretica* and *Salsola baryosma* and proper soil and water conservation measures.

10

ROLE OF REMOTE SENSING IN THE DETECTION OF TEMPORAL CHANGES

INTRODUCTION

In India various types of natural ecosystems such as fluvial, aeolian, marine, glacial and fluvio-glacial have been identified. The Land resources and land use patterns associated with these ecosystems are undergoing changes due to the natural and accelerated degradational processes. As a result, the ecological balance of the fragile ecosystems has been disturbed, resulting into the depletion of their biological productivity. The reliable and comprehensive information on the temporal changes in the land resources and land use is an essential pre-requisite for the proper planning and management of various ecosystems. Since the ecosystems are highly dynamic in nature and changes occur at variable rate from place to place, the detection of changes by conventional methods are time consuming and costly. Remote sensing techniques, particularly the satellite remote sensing due to its synoptic view of large area and spatial and spectral characteristics have been proved faster and economic in detecting the temporal changes in land resources and land use.

In the last few decades, the use of aerial and space remote sensing techniques has been made in detecting the changes in land use/land cover, forest area, landforms and water bodies etc. The results of these studies conducted in different parts of the country have been reviewed in this chapter.

CHANGES IN SALINE AREAS

It has been observed from the overlay of the maps on each other of saline areas prepared from the Landsat imagery of 1977 and Survey of India topographical map of 1960 that in the middle Luni basin near Sanwarla, Nosra, Samuja, Khutani, Chandari and between Harji and Manadar, large agricultural lands which were non-saline in 1960 have become saline in a period of 17 years i.e. 1960-77. In quantitative terms, it has been computed that in 1960 only 38.12 sq. km area was affected by salinity but in 1977, the salinity affected area increased to 103.60 sq. km, an increase of about 117 per cent (Singh and Shankarnarayan, 1983). This land degradation is indeed alarming from the view point of the decrease in productivity of the land (Fig. 10.1).

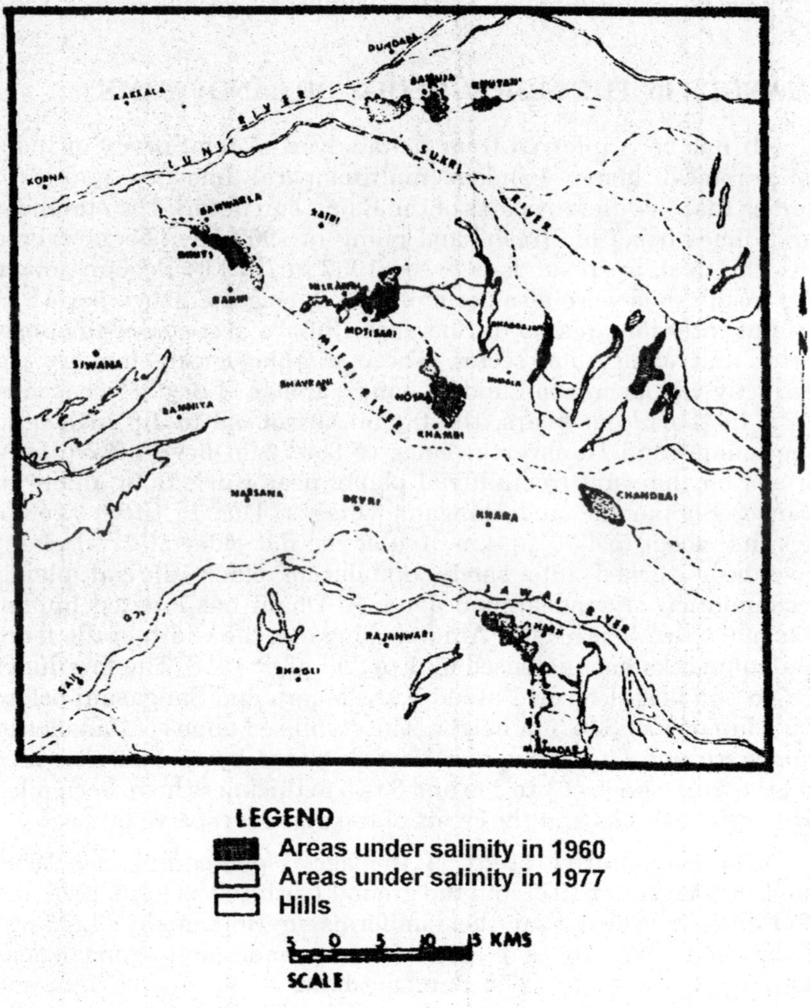

Fig. 10.1. Temporal changes in the saline areas in the south-eastern part of the middle Luni basin

Interpretation of the aerial photographs and field studies revealed that fertile agricultural lands around Nimbla and Molkasni, between Lolawas and Bhetanada in the above basin which were non-saline before 1958, have turned saline due to the impeded drainage conditions and rise of subsurface salts. As a result, a total area of 1560 sq. km has been affected by salinity hazards since 1958. It has also been observed that about 8.30 sq. km area of the non-saline younger alluvial plain along the Luni river right from Kankani to Shikarpura has become saline since 1958 (Singh *et al.*, 1978).

CHANGES IN THE MORPHOLOGY OF LANDFORMS

It has been inferred from the analysis of the Survey of India topographical maps, Landsat multitemporal imagery and field studies that in different parts of the Rajasthan desert, the stabilised sand dunes free from fresh sand piling in 1962 have been covered with the thick fresh sands between 1967 to 1979 by aeolian action. The fresh sands were also accumulated during the above period on the flat interdune plains in Shergarh-Chhaba sector, Siwana-Jasol sector and Jalor-Ahor sector. These morphological changes are distinctly visible on the Landsat band 7 image of dry hot season of 1977. In the recent years, the hummocks of 0.9 to 1.5 m height, longitudinal and transverse dunes of 3 to 6 m height have been formed on the sand free alluvial plains near Kuri, Doli, Jhanwar, Phinch, Bhetanada, and Gangana villages. Due to these aeolian deposits, about 524.30 sq. km area of the flat older alluvial plains have been turned into sandy undulating older alluvial plains. Accumulation of sand around Jaton-Ki-Dhani has created hummocks of 0.6 to 2.5 m height. Around Gangana, the width of the fence line hummocks has increased by 1 to 2 m after 1958. The low dunes of 1 to 2 m height formed aroud Kuri, Mogra and Sangasani before 1958, are now of 3 to 5 m height. The stabilised dunes of Janadesar, Chincharli and Uter have also been degrated by aeolian activities and the fresh sands of 1 to 2 m and 3 to 5 m thickness have been piled up along the flanks and the crests of the dunes, respectively.

The comparative study of the aerial photographs of 1958, Landsat imagery of 1977 and the ground truth collected in 1974 and 1981 have revealed that the landforms in Bhikamkaur-Lohawat sector and Tena-Balesar sector have undergone conspicuous morphological changes. The Barchan dunes of 3 to 6 m height and the shrub- coppice dunes of 1 to 2 m height have been created on the flat lands and in the bed of the stream near Balesar and Tena. The fresh sands of 1 to 2.5 m thickness near Lohawat have also been piled up on the flanks and the crests of the stabilised parabolic dunes. The fresh sands right from the leeward side of the parabolic dunes have moved horizontally to the adjoining sand free flat lands upto a distance of 38 m in last 20 years. These superficial deposits have depleted the productivity of the fertile agricultural lands.

CHANGES IN DRAINAGE SYSTEMS

In the Indian arid environment, the rivers are ephemeral in nature and carry water only as a direct response to torrential rainfall, resulting into flash floods. These flash floods appear suddenly and cause significant changes in the form of channel widening, deepening, bank erosion and deposition by bank scour and slumping. It has been detected from Landsat MSS data that due to the flash flood of 1979, the width of the Luni at Kankani village has increased from 298 m to 1500 m. The width of the Mitri at Pipar has increased from 70 m to 110 m, at Jaspali from 85 m to 160 m and at Benan from 70 m 140 m. The left bank of the Luni near Nimbla has been dissected by lateral shifting and its width has increased from 450 m to 1800 m.

The comparative studies of the Survey of India topographical maps of 1934, aerial photographs of 1970, and the Landsat imagery of 1979 revealed that near Ahor a new diversion channel of the Jawai river has been formed which passes through Ahor and Bhainswara and meets the Jawai again in the south of Sakrana. Near Punawas and Deta, the Sukri river ms sharp bend had made a new straight channel segment from Punawas to south of Golia (Singh and Ghose, 1977).

In the Rajasthan desert, the superimposition of maps of river courses and flood plains prepared from Survey of India topographical maps of 1958 and Landsat TM false colour composite of 1986 revealed that over a period of 28 years the river courses widened by 1.8 times and their areal extent increased due to 2-4 m dissections of the flood plains (Table 10.1). The non-degraded flood plain was largely valuable double cropped land. This fertile land due to changes in river courses was occupied by channel sand bars and point bars of 1.0 to 1.5m height, 1.5 to 2 m length and 0.8 to 1.2 m width and is having poor economic value. The flash flood overtopped the banks and deposited riverine sediments in the flood plains in the form of sand sheets of 0.2 to 0.8 m thickness, and sand mounds and sand ridges of 1.0 to 2.5 m height, 150 to 950 m length and 120 to 350 m width (Singh *et al.,* 1988).

Table 10.1 Temporal changes in the river courses and associated flood plains

River	Area (km^2)		Average width (m)	
	1958*	1986**	1958*	1986**
Luni	38.10	68.27	310	555
Mitri	8.50	15.53	121	222

*Based on Survey of India topographical maps.
** Based on Landsat TM false colour composite.
Source : Singh *et al.* (1988)

CHANGES IN WATER BODIES

An overlay of the maps of water bodies prepared from the Landsat TM data of 1986 and the Survey of India topographical maps of 1958 showed spectacular changes in the surface area of water bodies in the arid zone of Rajasthan (Table 10.2). Shankarnarayan and Singh (1979) have observed that the changes in the size and shape of water bodies are largely a function of rainfall occurrence. But in the present case, 1.8 to 2.4 times reduction in the drainage basin areas over a period of 28 years (1958-86) had resulted due to biotic interference like reclamation of drainage basin areas for cultivation and construction of houses, factories, etc. because the rainfall amount was almost same i.e. 254 mm and 237 mm during the years 1958 and 1986, respectively, with the similar distribution pattern (Ramakrishna, personal communication). This decrease in the drainage basin areas had also caused 6 to 8 times corresponding decrease in the surface areas of water bodies (Table 10.2). This also indicated that due to the interaction of man, land and water, desertic environmental conditions have been deteriorated, resulting into the desertification of large acreage of adjoining agricultural lands (Sharma *et al.* 1989).

Table 10.2. Temporal changes in the water surface and drainage basin areas of water bodies (*nadis*) in different landforms

Landforms	Water surface area (m^2)		Drainage basin area (ha)	
	1958	1986	1958	1986
Older alluvial plains	38620	7153	116.4	72.2
Interdunal plains	35000	4626	208.9	137.5
Rocky/gravelly pediments	63571	9169	527.7	231.9
Buried pediments	36636	9346	148.6	76.1

Source : Sharma *et al.* (1989).

Multidate Landsat imagery of cool, hot, and wet seasons revealed significant changes in the area of four water bodies (tanks) in the middle Luni basin of the western Rajasthan (Singh and Shankarnarayan, 1983).

The areas of all the tanks computed from the sequential image of one Landsat scene are given in Table 10.3.

Table 10.3. Changes in the surface area of the water bodies (tanks)

S. No.	Years and Scene ID	Area (km^2)			
		Nayagaon tank	Sardar samand tank	Kharda tank	Hemawas tank
1	2	3	4	5	6
	Cool Season				
1	Jan. 10, 1973 (1171-050-2)	0.24	4.23	0.85	1.95
2.	Feb. 14, 1975 (2023-04574)	5.62	5.29	2.84	0.39
3.	Jan. 16, 1977 (2725-04441)	5.62	45.90	17.39	12.55
	Hot Season				
4.	April 3, 1979 (2437-04532)	3.92	45.41	18.21	15.00
5.	March 11, 1977 (2779-04420)	4.01	98.12	14.17	4.32
6.	May 22, 1977 (2851-04382)	2.71	33.42	10.30	0.42
	Wet Season				
7.	Sep. 24, 1972 (1063-05090)	1.68	8.63	9.72	12.28
8.	Sep. 19, 1973 (1423-05081)	10.56	80.00	24.40	38.94

Source : Singh and Shankarnarayan (1983).

It is seen (Table 10.3) that in the cool season of January, 1973 and February, 1975, all the four tanks show appreciable shrinkage in the area because the preceding years 1972 and 1974 received low rainfall. The large size of these tanks in January, 1977 is attributed to the excessive rainfall received in 1976. In the hot season, the area under all the four tanks in May, 1977 is smaller than that of April, 1976 and March, 1977. It may be due to increased use of water for irrigation and high evapotranspiration. In the wet season, the tanks have larger area in September, 1973 than that of September, 1972 because the latter was a drought year.

CHANGES IN THE LAND USE

The study on detecting the temporal changes in land use of the middle Luni basin in the western Rajasthan revealed that significant changes in the acreage under irrigated and rainfed crops have occurred between 1973 and 1977, (Singh and Shankarnarayan, 1983). The comparative studies of Landsat false colour composites showed that 9% area under irrigated crops in January, 1973 increased to 12% in February, 1975. The increase was the highest 15% in January, 1977. The changes in the area under rainfed crops in September, 1972 increased to 80% in September, 1973 (Fig. 10.2).

It is inferred that Landsat false colour composites of cool and wet seasons are useful for detecting landuse changes which will form a basis for rational land use planning.

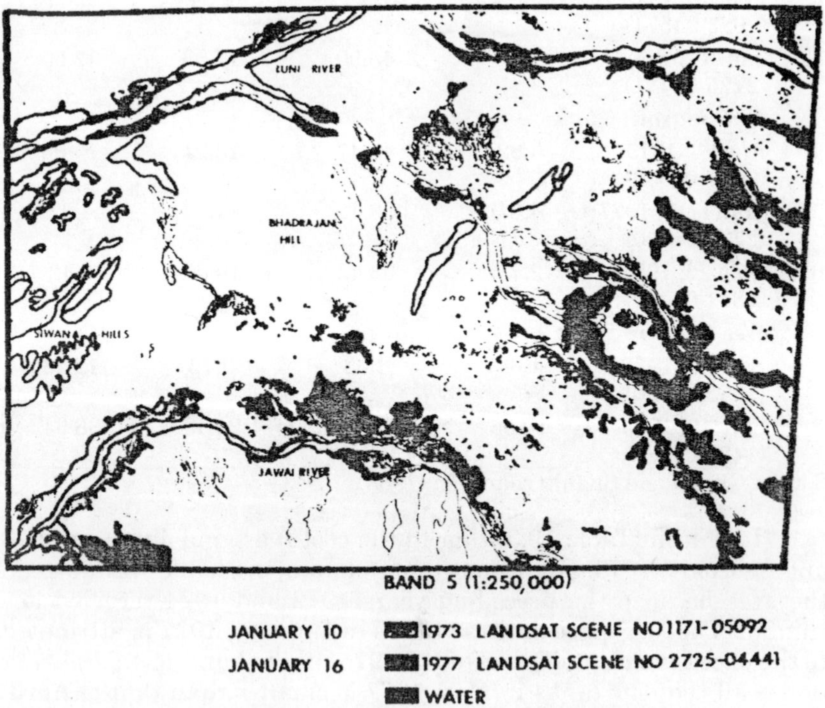

Fig. 10.2. Southeast segment of scene path 160 Row 042 showing changes in irrigated cropped area in 1973 and 1977

Role of remote sensing for mapping and monitoring changes in land use pattern in north-western part of Uttar Pradesh has been discussed (Sharma and Garg, 1986). The areal extent of the land use categories computed from the maps prepared from Landsat imagery are presented in Table 10.4 which also shows the land use changes over a period of 7 years (1974 to 1981).

Table 10.4. Changes in land use categories

S. No.	Categories	Areal extent				% change (1974-81)
		May 1974		May 1981		
		Area (ha)	(%)	Area (ha)	(%)	
1	Dry sand/soil	61376	10.96	66192	11.82	+0.86
2	Wet sand/soil	18368	3.28	14392	2.57	j0.71
3	Agricultural land	382368	68.28	389592	69.57	+1.29
4	Grassy/waste land	10416	1.86	8512	1.52	j0.34
5	Forests	85344	15.24	78960	14.10	j1.14
6	Built-up area	2128	0.38	2352	0.42	+0.04
	Total	5.6x10^{5}	100	5.6x10^{5}	100	j

Source : Sharma and Garg (1986).

The results indicate that forests have decreased by 1.14% indicating deforestation for firewood and various other human needs. Agricultural land has shown an increase of about 1.3% and grassy/wasteland has decreased by 0.34%. Built-up area has increased only 0.04% which shows slow but gradual urbanisation in this predominantly agricultural region.

CHANGES IN GRAZING LANDS AND FOREST COVER

The superimposition of maps of grazing lands prepared from the Survey of India topographical maps of 1958 and Landsat TM false colour composite of 1986 revealed that in the Rajasthan desert over a period of 28 years, the areal extent of grazing lands decreased by 9 to 30% under various landforms (Table 10.5). The decrease in the areal extent of grazing lands associated with the saline alluvial plain is lowest due to the salinity hazard and highest in case of rocky/gravelly pediment because much of the land has been reclaimed for cultivation due to the availability of sweet ground water. The decrease in the areal extent of grazing lands is due to the increased human activity like cultivation of marginal lands, construction of residential dwellings and industries on them and the increasing trend of urbanisation (Sharma *et al.*, 1989).

Table 10.5. Changes in areal extent of grazing lands under different landforms

Landforms	Extent of grazing lands (ha)		Reduction in the area (%)	No. of samples
	1958*	1986**		
Sandy alluvial plain	316	223	29	59
Younger alluvial plain	154	125	18	26
Saline alluvial plain	170	155	9	13
Rocky/gravelly pediment	527	367	30	25

* Based on Survey of India topographical maps.
** Based on Landsat TM false colour composite.
Source : Sharma *et al.* (1989).

The temporal changes occurred during the years 1975 and 1983 in forest cover over a period of 8 years in the part of Gorakhpur district, Uttar Pradesh were detected by using Landsat multi-spectral data (Kachhwaha, 1985). The results of this study revealed that there is only 0.24% depletion in the area under forest from 1975 to 1983 but the quality of the forest in terms of crown density has reduced to a considerable extent. High density dominated forests have reduced by 28.72 sq. km during a period of 8 years. This depletion in density forests area is 7% of the total forest area.

Based on the visual interpretation of Landsat false colour composite on 1 : 1 million scale and selected field checks for two periods 1972-75 and 1980-82, the changes detected in forest area in the Rajasthan are presented in Table 10.6 (NRSA, 1983).

During period from 1972-75 to 1980-82 over all reduction of 1.55% of the total geographical area has occurred. The closed forest area has reduced by 51.4 sq. km (1.49%) and open/degraded forest area by 218 sq. km (0.06%). The reduction in total forest area over the period of 7 years is 5322 sq. km (47.12%). This study has indicated that satellite remote sensing can be valuable tool to detect the changes in forest areas in a relatively short time span compared to conventional methods.

Table 10.6. Area under different forest classes in Rajasthan during 1972-1975 and 1980-1982 (based on visual interpretation of Landsat data)

Period	Area (km^2)				Percent forest area of total geographical area
	Total geographical area	Closed forest	Open degraded forest	Total forest area	
1972-75	342210	8364	2930	11294	3.30
1980-82	k	3260	2712	5972	1.75
Changes in forest area		5104 (1.09%)	288 (0.06%)	5322 (47.12%)	1.55

Source : NRSA (1983)

Digital temporal analysis was done by Ramesh *et al.* (1986) to monitor the dynamics of vegetation in small watershed over a period of 10 years (1973-1984). Normalised difference vegetation index (ND) was evaluated for estimating the vegetation cover changes over these years. The results of this study show that there is general increase in vegetation density in the upper reaches of Mulaiar and Iruttar micro watersheds due to increase in the plantation activity and afforestation especially around the areas of Valagiri, Perumalai and Palamalai.

It is inferred from this study that the temporal changes in landforms, landuse/land cover, water bodies and saline areas etc. could be detected easily, rapidly and accurately by aerial and satellite remote sensing techniques than by conventional methods. Based on the mapping of the above temporal changes over a space and time, suitable measures could be adopted to control these changes.

11

APPLICATION OF REMOTE SENSING IN WASTELANDS MAPPING

INTRODUCTION

In India out of total geographical area of 329 m ha, large acreage of land has been turned into wastelands due to several natural and biotic factors like drought, water erosion, wind erosion/ deposition, salinity/alkalinity, floods and un-scientific methods of cultivation. The studies conducted by several workers on the type and distribution of wastelands by using conventional methods are costly and time consuming and do not provide reliable information on their distribution and extent.

Realising the importance of the systematic classification and mapping of wastelands for their rational development to upgrade the production of food, fodder and fuel, the National Remote Sensing Agency, Hyderabad (1985) completed the wastelands mapping of each state of India on 1 : 1 million scale by using Landsat false colour composites of the same scale and random field checks. These maps were not found suitable to plan proper measures for development of wastelands. But they provided the guide lines to the Technical Task Group constituted by Planning Commission in 1986 in finalisation of the definition and categories of wastelands and in fixing the priorities for the mapping of wastelands on 1 : 50,000 scale at district level in different states of India by using Landsat enlarged Thematic Mapper (TM), False Colour Composites (FCCs) of the same scale.

Based on the satellite remote sensing techniques and ground truth, the type, distribution and extent of wastelands in India in general and in Jodhpur District of western Rajasthan in particular are discussed.

TYPE, EXTENT AND DISTRIBUTION OF WASTELANDS IN INDIA

Different types and definitions of wastelands have been proposed by various workers. Technical Task Group constituted by Planning Commission. Govt. of India identified 13 categories of wastelands. As defined by the Technical Task Group, *the wasteland is that land which is degraded and is presently lying un-utilised except current fallows due to different physical constraints* (1986). But this definition was not found suitable for the development of wastelands and an Interministerial Working Group was, therefore, constituted by Prof. M.G.K. Menon.

According to this Working Group, *the wastelands mean degraded land which can be brought under vegetative cover with reasonable effort and which is currently under-utilised and land which is deteriorating for lack of proper water and soil management or on account of natural causes. Wasteland can result from inherent/ imposed disabilities such as by location, environment, chemical and physical properties of the soil or financial or management constraints* (NWDB, 1987).

It has been reported by NRSA (1985) that in India out of total 53.30 million ha of wastelands, 32.85 million ha or 61.5% is under culturable wastelands and 20.45 million ha or 38.5% under un-culturable wastelands. The greatest extent of wasteland in the country is in Rajasthan 12.94 million ha, followed by Uttar Pradesh 4.31 million ha and Gujarat 3.3 million ha.

Based on the visual interpretation of Landsat data on 1:1000,000 scale, the salient characteristics of the following major types of wastelands in India have been highlighted by Gautam and Narayan (1988).

Salt Affected Land

It has been estimated that 3.9 million ha of land is salt affected. The highest concentration of this type of wasteland is in Gujarat, followed by Uttar Pradesh and Punjab. This category of wasteland in the States like Himachal Pradesh, Jammu and Kashmir, Assam, Tripura, Manipur, Mizoram, Arunachal Pradesh, Meghalaya and Nagaland is either totally absent or is concentrated in small pockets due to hilly terrain. These wastelands appear in white to light blue tone and fine to mottled texture on Landsat false colour composites masaic of Rajasthan and adjoining regions (Fig. 11.1).

Gullied or Ravinous Lands

This category of wasteland comprising of an area of 4.3 million ha is largely associated with the rivers like the Yamuna, the Chambal, the Tapi,the Krishna and the Tungabhadra etc. The gullied/ravinous areas could be easily separated and mapped from the interfluve areas on Landsat false colour composite image.

Gullied lands appear in light yellow to bluish green tone whereas the interfluve areas due to the stand of crops in wet season exhibit red tone on above Landsat image (Fig. 11.1).

Waterlogged or Marshy Land

This category of wasteland constitutes 0.9 million ha, of which 75% is in West Bengal, Uttar Pradesh and Bihar. Impeded drainage conditions, high water table, seepage from canal and reservoirs and presence of impervious layer at shallow depth are responsible for the development of this type of wasteland. Waterlogged or marshy areas in different shape and size are distinctly visible on Landsat false colour composites mosaic of Rajasthan and adjoining regions (Fig. 11.1).

Upland With or Without Scrub

This is the most dominant category of wasteland which occurs throughout the country and constitutes 11.8 million ha of land. It largely occurs in Maharashtra, followed by Andhra Pradesh, Karnataka, Madhya Pradesh and Rajasthan. These wastelands appear in light yellow to dark yellow tone with red patches on Landsat false colour composites mosaic of Rajasthan and adjoining regions (Fig. 11.1). In summer season, it becomes difficult to separate the fallow lands from uplands due to similar reflectance. In wet season when the fallow lands are put under cultivation, the uplands with or without scrub could be easily separated and mapped from fallow lands.

Jhum or Forest Blanks

This category of wasteland, by and large, occurs in the north-eastern region comprising of Assam, Arunachal Pradesh, Mizoram, Manipur, Meghalaya, Nagaland and Tripura. The total area under this category is 2.4 million ha and Manipur has the largest area, followed by Madhya Pradesh, Assam and Mizoram etc. It is easy to distinguish forest blanks amidst forest on Landsat false colour composites mosaic due to their white tone, shape and size, whereas the forest areas appear in red tone.

Sands (Coastal or Desertic)

This is second largest category of wasteland which comprises about 10.5 million ha of land. The coastal sands lie along the eastern and western coast of Peninsular India, whereas the desertic sands occur in north-western parts of the country. The maximum area under desertic sands is concentrated in western Rajasthan, followed by Punjab and Haryana. Both coastal and desertic sands due to their location, shape, size and patterns are distinctly visible on Landsat false colour composites mosaic of Rajasthan and adjoining regions in white to grey tone with red patches. (Fig. 11.1).

TYPE, EXTENT AND DISTRIBUTION OF WASTELANDS IN JODHPUR DISTRICT

In Jodhpur district, out of 13 wasteland categories suggested by the Technical Task Group, only 8 categories have been identified (Singh *et al.,* 1989). As an example, some categories have been mapped (Fig. 11.2). The extent of each category is given in Table 11.1.

Gullied and/or Ravinous Lands

This category of wasteland is associated with graded river beds, younger alluvial plains, hill, rocky/gravelly and buried pediments (Singh, 1983). The depth and width of the gullies ranges from 3 to 8m and 2.5 to 5.3m, respectively. Some of the hill slopes covered with aeolian sands have been dissected by water action into 10 to 25m deep and 30 to 50m wide gullies, resulting into *bad land topography*. The soil texture along the gullies ranges from sand to gravelly sand and loamy sand to sandy loam. The beds of the gullies are flat and covered with gravelly sands and gravels. It covers 0.40% of the total geographical area and 1.30% of total area of wastelands of the district. The gullied lands appear in white to grey and light yellow tone on Landsat TM false colour composite and could be easily identified due to their shape, size and pattern. Such wastelands exhibit white to grey and grey to dark grey tone on Landsat band 7 black and white image (Fig. 11.3).

Table 11.1. Area and percentage to total of each category of wasteland

	Category	Area of wasteland (km^2)	% to total geographical area	% to total area of wastelands
1	Gullied and/or ravinous lands	91.46	0.40	1.30
2	Upland with or without scrub	2197.20	9.62	31.47
3	Land affected by salinity/alkalinity	169.22	0.74	2.44
4	Under-utilised/degraded notified forest lands	34.93	0.15	0.50
5	Degraded pastures/grazing lands	1267.55	5.55	18.16
6	Sand-desertic	3103.77	13.58	44.46
7	Mining/industrial wastelands	25.63	0.11	0.37
8	Barren rocky/stony waste/sheet rock areas	90.82	0.40	1.30
		6980.98	30.55	100.00

Source : Singh *et al.* (1989).

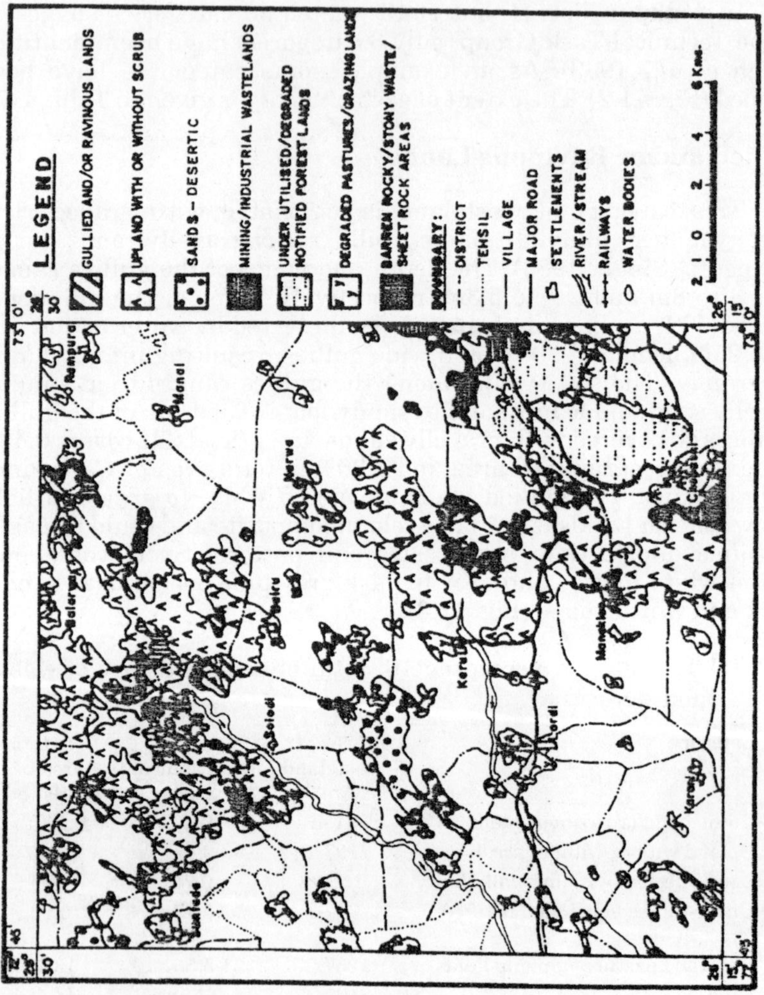

Fig. 11.2. Jodhpur district - Wasteland types

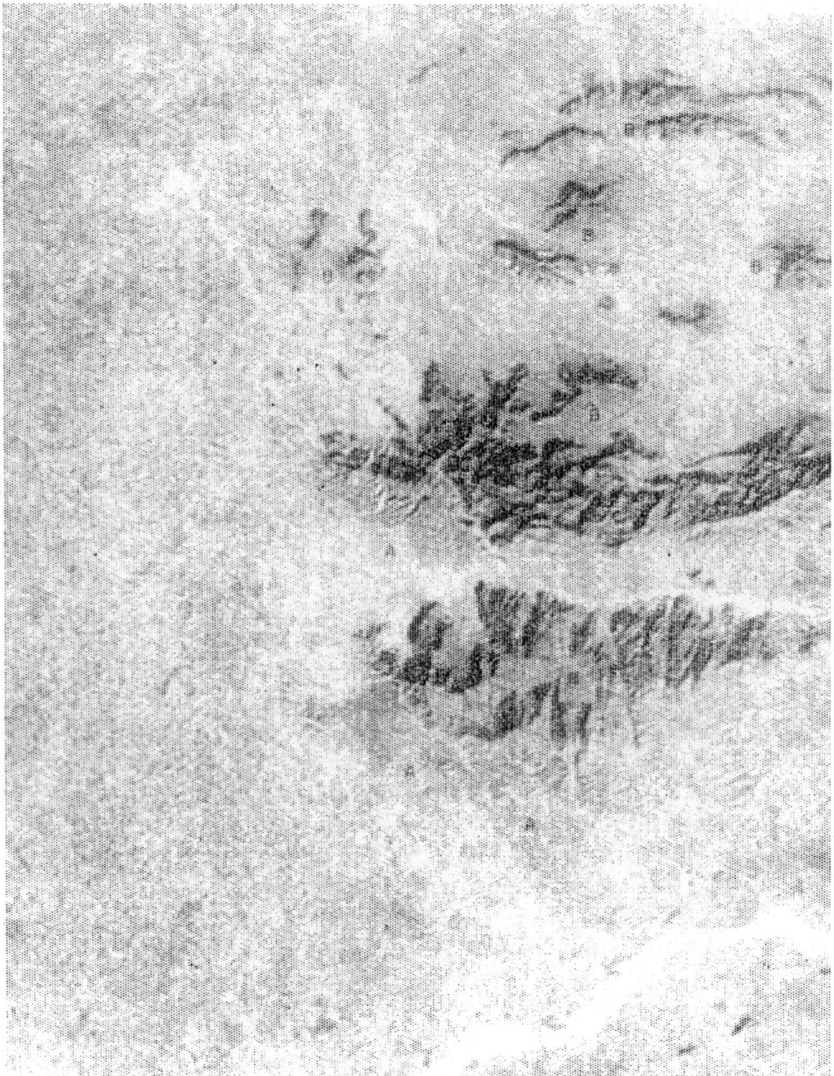

Fig. 11.3. Landsat band 7 black and white image showing spatial and spectral characteristics of gullied and/or ravinous lands (A) and rocky barren/ stony waste and sheet rock areas (B)

Upland With or Without Scrub

This type of wasteland associated with rocky gravelly pediments, plateaus and rock outcrops occurs in all the tehsils of the district (Singh, 1983). It covers 9.62% of the total geographical area and 21.47% of total area of wastelands of the district. The salient physical characteristics are rocky, gravelly and stony surface, flat topography interrupted by rock outcrops, shallow deposits in pockets and presence of drainage channels along the joints and fractures. These wastelands appear in grey to dull white and dark brown tone on Landsat TM false colour composites. This category of wasteland distinctly appears on Landsat band 5 black and white image in white to grey and dark grey tone (Fig. 11.4).

Land Affected by Salinity/Alkalinity

This wasteland category covers 0.74% of total geographical area and 2.44% of total area of wastelands of the district. This category constitutes two types of wastelands viz., natural and man-induced. Natural saline/alkaline wastelands are genetically related to the disorganised courses of the streams and are located at their confluences. The sand to loamy sand and loam to clay loam soils are highly saline with pH 7.8 to 8.6 and EC 1.82 to 29.30mmhos/cm. This category of wasteland is used for extracting salts. Man-induced saline/alkaline wasteland have developed due to the excessive use of highly sodic waters for irrigation and rise of ground water table and construction of embankments across the buried drainage channels (Singh *et al.* 1978). The soils are highly saline with pH 8.3 to 9.6 and EC 0.4 to 6.3 mmhos/cm. The salt encrusted parts of this wasteland category exhibit white to grey tone and the areas filled with water appear in light blue tone on Landsat TM false colour composite. Such wastelands due to variations in degree of salinity appear in bright to grey and dark gray tone with mottled texture on Landsat band 5 black and white image. (Fig. 11.5).

Under-utilized/Degraded Notified Forest Lands

This type of wasteland occupies 0.16% of the total geographichal area and 0.50% of total wasteland of the district. It is characterised by rocky, gravelly and stony surfaces and pebbles and gravels mixed with loamy sand to sandy loam soils of 20 to 25 cm depth. The sheet, rill and gully erosion are the major physical limitations of this wasteland. This category of wasteland also occurs on the light to medium and medium to heavy textured alluvial plains in Bilara and Bhopalgarh tehsils. These wastelands appear in yellowish green and light red tone on Landsat TM false colour composite and could be easily identified due to their distinct tonal and spectral contrast to the surrounding areas. This category of wasteland exhibit white and dark grey tone on Landsat band 5 black and white image. (Fig. 11.4).

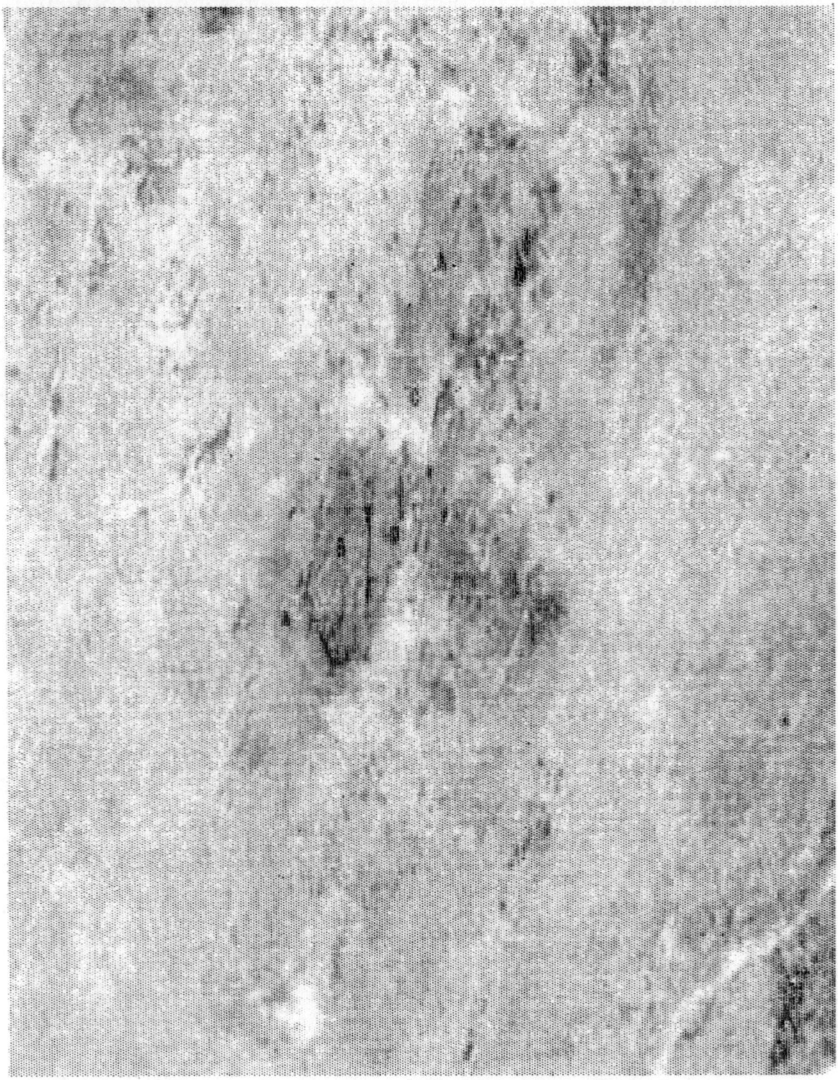

Fig. 11.4. Landsat band 5 black and white image exhibiting spatial and spectral characteristics of uplands with or without scrubs (A); under utilized/degraded notified forest land (B) and mining wastelands

Degraded Pastures/Grazing Lands

This category of wasteland occupies 5.55% of the total geographical area and 18.16% of total wasteland and occurs in all the tehsils of the district. The flat topography, gravelly surfaces, sand to loamy sand and sandy loam to loam and clay loam soils, salinity and water erosion are the major physical characteristics of this wasteland. These wastelands appear in white, milky white, whitish grey and black grey tone on Landsat band 5 black and white image (Fig. 11.5) and in light red tone on Landsat TM false colour composite.

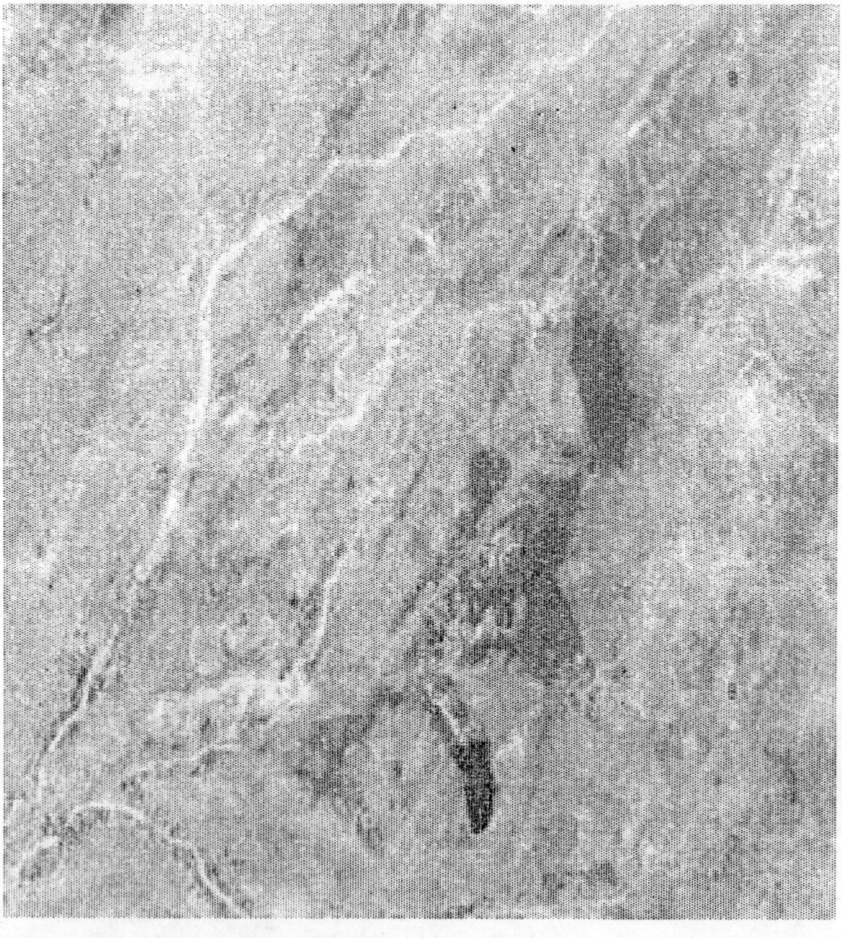

Fig. 11.5. Landsat band 5 black and white image showing spatial and spectral characteristics of lands affected by salinity/alkalinity (A) and also areas occupied by degraded pastures/grazing lands (B)

Sands (Desertic)

This category of wasteland occupies the largest area, 13.58% of total geographical area and 44.4% of total wasteland of the district. The stabilised and unstabilised sand dunes are the major constituents of this category of wasteland. Sandy wastelands comprising of stablised parabolic, longitudinal and transverse sand dunes are characterised by high relief of 5 to 70 m, gently to moderately steep slopes, compact and loose structureless sands, poor fertility and water holding capacity of soils and presence of alternate layers of calcium carbonate and high vulnerability to wind erosion. The stabilised sands with vegetation appear in light bluish green to red tone and the fresh sands exhibit white tone on Landsat TM false colour composite. Due to distinct shape and size of sandy deposits, these wastelands in the form of coalesced parabolic, longitudinal and transverse sand dunes distinctly appear in white to grey and dark grey to dark tone on Landsat band 7 black and white image (Fig. 11.6).

Mining/Industrial Wastelands

It constitutes 0.11% of total geographical area and 0.37% of total wasteland of the distinct. Due to mining activities large areas have been spoiled and turned into *bad lands*. The mining waste in the form of the heaps of debris and detritus poses the problem of wind erosion. Water erosion in the form of rills and gullies is another problem of this wasteland. Industrial wasteland has developed along the Jojri river due to the industrial effuent from the textile industries located in Basni area. The hard soil crust, poor fertility, stunted growth of the crops and thick deposits of effluent are the salient characteristics of this type of wasteland. Mining areas appear in white to grey tone and areas affected by industrial effluent exhibit light blue tone on Landsat TM false colour composite. The sandstone mining areas near Sursagar due to milky white and grey tone could be easily identified from Landsat band 5 black and white image. (Fig. 11.4)

Barren Rocky/Stony Waste and Sheet Rock Areas

This type of wasteland occupies 0.40% of the total geographical area and 1.30% of total wasteland of the district. The barren hills comprising of sandstone, granite and rhyolite rocks and sheet rock form this type of wasteland. The steep slopes devoid of debris, flat top, sealed rocky surfaces and water erosion are the major physical

Fig. 11.6. Landsat band 7 black and white image showing form, pattern, spatial and spectral characteristics of sandy wastelands

constraints which do not allow the vegetation to grow on this wasteland. This category of wasteland appears in white to grey and black tone on Landsat band 7 black and white image (Fig. 11.3) and in grey to black tone with red patches on TM false colour composite.

DEVELOPMENT OF WASTELANDS

Wastelands identified in the district could be developed according to their landuse capability. The arable farming due to the limitations of biophysical potentials of the wastelands is not economically viable. The wastelands like sands (desertic), gullied and/or ravinous lands, uplands with or without scrub could be developed into pastues, silvipastures and woodlands by using proper soil and water conservation measures like wind breaks, shelterbelts, contour furrows and contour bunding. The indigenous plant species like *Prosopis cineraria, Prosopis julifora, Acacia senegal, Salvadora oleoides, Capparis decidua, Calligonum polygonoides, Ziziphus nummularia, Lasiurus sindicus, Cenchrus ciliaris and Panicum turgidum* can very well thrive on these wastelands.

The fodder produced from the above plants could be used to develop animal husbandry with best breed of cattle, goat and sheep which will be an additional source of income to the rural people and will improve their socio-economic conditions. The wood available from the plants may be used for fuel and to make agricultural implements. The productivity of degraded pastures/grazing lands could be increased by adopting deferred rotational grazing and by introducing high yielding varieties of trees, shrubs and grasses. The fruit plants like ber (*Ziziphus mauritiana*), pomegranate (*Punica granatium*), custard apple (*Annona squamosa*) and lasoora (*Cordia myxa*) could be established on the wastelands (Dass, 1986). These fruit crops will become a permanent source of income to the rural people and some fruit crops in the form of fuel wood and fodder will provide additional income.

The wastelands like uplands with or without scrub and barren rocky/stony waste/rock sheet areas could be developed into small catchments and surface runoff could be harvested at suitable sites for drinking and pond farming which will provide additional income to the villagers. The parts of these wastelands, where mining activities have not been developed so far, could be systematically mined. The mining products like sandstone, limestone and gypsum will help in the construction of roads and buildings for houses, schools, health centres, cooperative banks, grain and fodder storage. The development of mining activities on wastelands will provide employment to the rural people and an additional source of income.

Based on the plants products like gum, timber, wood, fibre, pulp, rasin, oil, fruits and sajji from the wastelands, the agro-

Table 11.2 Wasteland types and agro-industries

S. No.	Wasteland types	Plant species	Local name	Plant products	Agro-industries
1.	Gullied and/or ravinous lands	*Acacia nilotica*	Desi babool	Gum, timber,	Gum, agricultural implements,
		Acacia tortilis	Israeli babool	Gum, softtwig	Gum, baskets, Tanin,
		Cassia auriculata	Anwal	Rasin	Soap,
		Salvadora oleoides	Jal	Oil, soft twig	basket
2.	Uplands with or without scrub	*Acacia tortilis*	Israel babool	Gum, wood	Gum, furniture,
		Commiphora wightii	Guggal	Oleo, rasin	Medicine, perfume
3.	Land affected by salinity/ alkalinity	*Suaeda fruticosa*	Kala lana	Sajji	Salt
		Haloxylon salicornicum	Jiri lana		
4	Under utilised/ degraded notified forest lands	*Acacia tortilis*	Israeli babool	Gum, wood	Gum, furniture
		Acacia nilotica	Desi babool	Gum, wood	Gum, furniture
		Acacia leucophloea	Rheonja	Gum, rasin	Gum, tanin
		Albizzia lebbek	Siris	Timber	Furniture
5.	Degraded pastures/ grazing lands	*Prosopis cineraria*	Khejari	Fruits	Dry vegetable
		Tecomella undulata	Rohida	Timber	Furniture
		Salvadora oleoides	Jal	Oil	Soap
		Ziziphus nummularia	Bordi	Fruits	Medicines jams
		Capparis decidua	Kair	Fruits	Pickles, medicines
6.	Sands-desertic	*Tecomella undulata*	Rohida	Timber	Furniture
		Prosopis cineraria	Khejari	Fruits	Dry vegetable
		Letadenia pyrotechnica	Khimp	Fibre	Rope
		Saccharum begalensis	Munja	Fibre, pulp	Rope, paper
		Citrullus colocynthis	Tumba	Oil	Soap
7.	Mining/industrial wastelands	*Acacia tortilis*	Israeli babool	Gum, wood	Gum, furniture
		Salvadora oleoides	Jal	Oil	Soap
		Capparis decidua	Kair	Dry fruits	Pickles

Source : Singh *et al.* (1989).

industries like soap, gum, basket, furniture, pickel, rope, paper, tanin, agricultural implements, mat and flour could be established (Table 11.2). These agro-industries will provide employment to the people and will be a permanent source of income to transform the rural economy.

It has been concluded from this study that remote sensing using aerial photographs and satellite imagery in conjuction with ground truth plays an important role in the delineation and mapping of various categories of wasteland more quickly, accurately and economically than by conventional methods. Each wasteland category could be developed according to its physical potentials and limitations.

12

DISTRIBUTION AND MANAGEMENT OF SAND DUNES USING REMOTE SENSING TECHNIQUES

INTRODUCTION

The Indian desert forming the part of hot arid zone of India constitutes 0.32 million km^2 area. The bulk of the area (61 per cent) of the desert lies in Rajasthan, followed by Gujarat (19.6 per cent); Andhra Pradesh (7 per cent); Punjab (5 per cent) Haryana (4 per cent); Karnataka (3 per cent) and Maharashtra (0.4 per cent).

Like other deserts of the world, sand dunes are the most spectacular and dynamic features of the great Indian desert. Dust storms and sand storms are common and the frequency of their occurrence is highest during the months of May and June when wind speed is more than 40 km/hr. Most common dune types identified by using remote sensing techniques and ground truth in the Indian desert are obstacle, parabolic, longitudinal, transverse, barchan and shrub-coppice. (Singh, 1982; Singh and Shankarnarayan, 1986).

In this chapter; the evolution, distribution and morphology of different dune types and the suitable techniques for their management and stabilisation have been discussed.

DISTRIBUTION AND MORPHOLOGY OF SAND DUNES

The distribution and morphological, vegetation and Landsat spectral characteristics of obstacle, parabolic, longitudinal, transverse and shrub-coppice sand dunes (Fig. 12.1) are discussed below.

Obstacle Sand Dunes

These sand dunes largely occur in the central and north-eastern parts of the Rajasthan desert. The windward and leeward

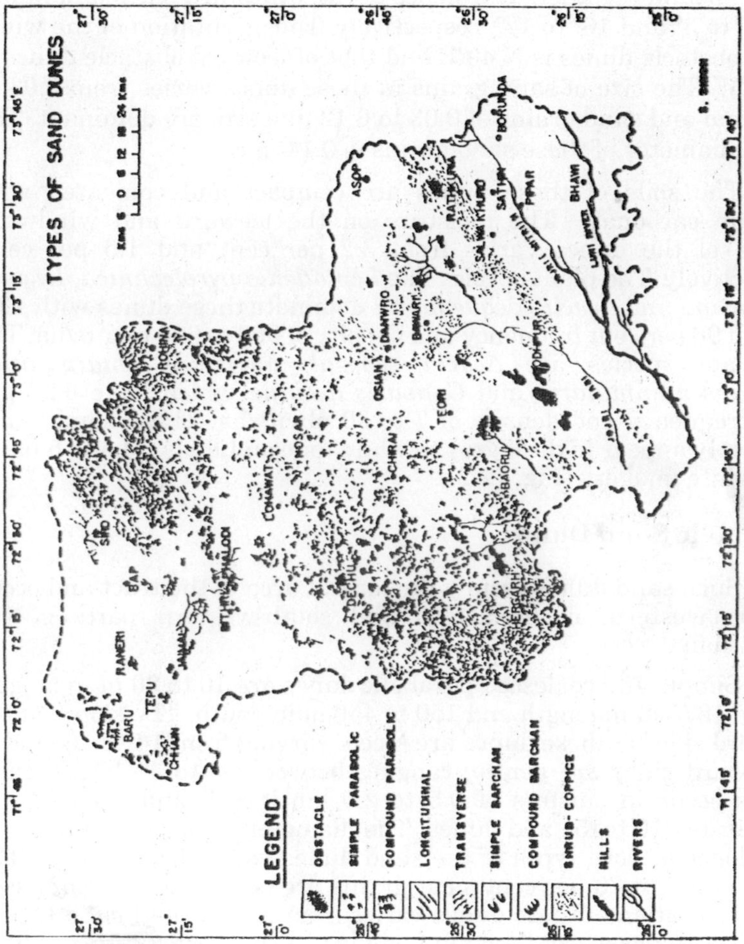

Fig. 12.1. Distribution of different types of sand dunes in Jodhpur district

obstacle dunes formed in the windward and leeward sides of the obstacles have been encountered in Jodhpur, Nagaur, Jaisalmer and Jalor districts.

The obstacle dunes are of 8 to 40 m in height, 200 to 450 m in length and 150 to 350 m in width. The alternate layers of calcium carbonate have formed within these dunes which keep the sand grains bound together. The slopes of the windward sides and flanks are 2° to 3° and 10° to 12°, respectively. The orientation of the windward obstacle dunes is N 40°E and that of leeward obstacle dunes is N 48°W. The size of sand grains in these dunes varies from 0.03 to 0.25 mm and sand grains of 0.03 to 0.12 mm size are dominant. The mean diameter of these sand grains is 0.14 mm.

The soils of these dunes are compact and cemented with calcium carbonate. The moisture on the leeward and windward slopes of the dunes varies from 3.2 per cent and 1.5 per cent, respectively. The plant species like *Leptadenia pyrotechnica, Lycium barbarum*, and *Acacia jacquemontii* dominate these dunes with 100, 46 and 96 per cent frequency and density of 360 to 620 plants/ha. The associate species are *Acacia senegal, Maytenus emarginatus, Ziziphus nummularia* and *Capparis decidua* which have 8.23 per cent frequency and density of 7 to 30 plants/ha. These sand dunes distinctly appear in dark grey to white tone on Landsat band 5 black and white imagery (Fig. 12.2).

Parabolic Sand Dunes

Such sand dunes occupy the largest area of the tract and occur in the western, north-eastern and south-western parts of the Rajasthan desert.

Simple and coalesced parabolic dunes are 10 to 20 m in height, 275 to 1275 m in length and 150 to 450 m in width. The slopes of the leeward sides of these dunes are steep, varying from 14° to 16° and of windward sides are gentle, ranging between 2° to 4°. These sand dunes occur in clusters of 0.5 to 2.5 km length and each cluster constitutes 10 to15 sand dunes. The diameter of the sand grains on the slopes of both types of the sand dunes varies from 0.03 to 0.25 mm and 60 to 70 per cent sand grains are of 0.06 to 0.12 mm size. The orientation of these dunes varies from N 45°E and N 50°E. The slopes of the leeward, flank and windward sides are 22° to 24°; 8° to 12° and 2° to 4°, respectively.

The soils on the surface of the dunes are loose and their compactness increases with the increase in depth. The water holding capacity of the soils varies from 22 to 25 per cent. The moisture at windward sides of dunes is least, 1.7 per cent, followed by flanks 2.5 per cent and crest 3.3 per cent. The dominant plant community of

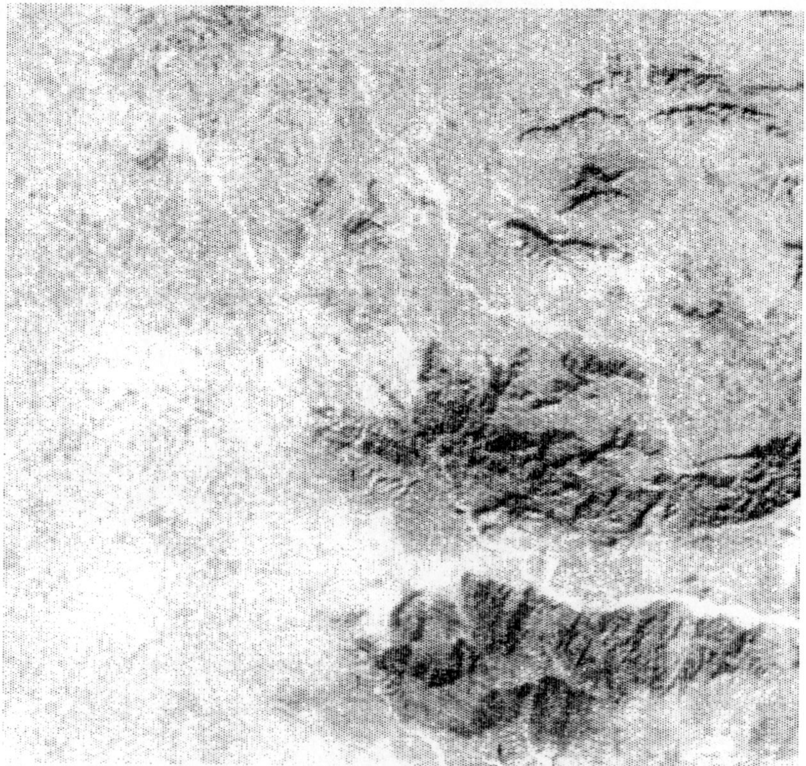

Fig. 12.2. Landsat band 5 black and white image showing the tonal, spectral and spatial characteristics of obstacle sand dunes

these dunes is *Acacia senegal, Calligonum plygonoides* with 58 to 78 per cent frequency and density of 150 to 187 plants/ha. The associate species are *Acacia jacquemontii, Ziziphus nummularia* and *Prosopis cineraria* with 20 to 40 plants/ha density. These dunes distinctly appear in grey to dark grey and dark tone on Landsat band 7 imagery (Fig. 12.3).

Longitudinal Sand Dunes

These sand dunes largely occur in south-western parts of Jaisalmer district and north-western parts of Jodhpur district, particularly in Phalodi tehsil.

The sand dunes of this type are well stabilised and their height varies from 15 to 60 m. The orientation of such dunes is in the direction of the prevailing south-west winds and it varies from N 40°E to N 45°E. The slopes of the leeward, flanks and windward sides are 20° to 24°, 9° to 14° and 2° to 3°, respectively. They are 1200 to 5000 m in length and 150 to 450 m in width. The sand grains of 0.07

to 0.15 mm diameter are dominant. The average diameter of sand grains on different dune slopes varies from 0.10 to 0.14 mm.

In these dunes, the soils have more than 85 per cent fine sand and rest are coarse sand, clay and silt. The moisture content within such dunes varies from 2.4 per cent on the windward slopes to 2.9 per cent on the crest and leeward slopes. They are dominated by trees and shrubs like *Acacia senegal, Salvadora oleoides, Acacia jacquemontii, Capparis decidua* and *Calligonum polygonoides* with 50 to 90 per cent frequency and 91 to 150 plants/ha density. The associate species are *Lycium barbarum, Calotropis procera* and *Maytenus emarginatus* with 31 to 46 per cent frequency and 20 to 57 plants/ha density. Such dunes due to their linear shapes distinctly stand out on Landsat black and white imagery of band 7 and they exhibit grey to dark grey and dark tone (Fig. 12.3).

Fig. 12.3. Landsat band 7 white and black image showing tonal, spectral and spatial characteristics of different type of sand dunes

Transverse Sand Dunes

These sand dunes in the form of parallel ridges are developed to the right angle of prevailing winds and largely occur in the eastern parts of Jodhpur and Nagaur districts. The high density of such dunes has been found in Pugal-Kanasar-Manaksar sector of Bikaner district.

The height of these dunes varies from 5 to 40 m. They are 1025 to 2560 m in length and 256 to 375 m in width. The orientation of the high dunes is N 50°W and that of low dunes is N 45°W. The slopes of the leeward, flanks and windward sides of the dunes are 20° to 22°, 12° to 14° and 3° to 4°; respectively. The size of sand grains of the dunes varies from 0.03 to 18 mm and about 50 per cent sand grains are of 0.03 to 0.07 mm diameter. The average diameter of the sand grains is 0.11 mm.

The soils of these dunes are fine sands through out the profiles. The moisture within such dunes varies from 2.4 per cent on the wind ward slopes and 3.6 per cent on the crest and the leeward slopes. The common plant species found on these dunes are *Calligonum polygonoides, Acacia jacquemontii, Panicum turgidum, Haloxylon salicornicum* and *Aristida funiculata*. Landsat band 7 black and white image facilitated to identify the transverse sand dunes which appear in white to grey and dark grey dark to dark grey tones and wavy patterns on this image (Fig. 12.3).

Barchan Sand Dunes

These dunes are simple and compound in nature and occur to the west of 300 mm isohyet near the settlements, on the flat lands and in the river beds. These sand dunes are 2 to 12 m in height, 100 to 350 m in width and 150 to 550 m in length. The coalesced dunes occur in clusters and each cluster comprises 5 to 15 dunes and spacing between them varies from 100 to 300 m. The orientation varies from N 38°E to N 40°E. The slopes of the leeward, flanks and widward sides of the dunes are 14° to 16°, 5° to 8° and 2° to 3°; respectively. The size of the sand grains ranges between 0.03 to 0.25 mm and the average size of the sand grains is 0.11 mm.

The soils of the dunes are fine grained, non-calcareous and loose. The moisture at 20 to 25 cm depth varies from 2.75 to 3.15 per cent and it increases with depth on all slopes of the dunes. Within these dunes, more than 5 per cent moisture prevails throughout the year at 1.5m depth (Krishnan *et al.,* 1966). These sand dunes are characterised by complete absence of vegetation but some plants like

Cyperus arenarius, Aerva persica, Citrullus colocynthis and *Tribulus terrestris* do appear between the extending arms of these dunes. These sand dunes occurring in the corridors of the longitudinal dunes in the extreme southern part of the Landsat band 7 black and white image appear in grey to dark grey tone (Fig. 12.3).

Shrub-Coppice Sand Dunes

These dunes are widely distributed in all parts of the Rajasthan desert and occur in the sandy undulating interdune plains and older alluvial plains. Such dunes are of different size, shape and orientation. The sand mounds are of 0.9 to 1.5 m height and their length and width are almost equal. They do not have any definite orientation and lie quite close to each other. The fence line hummocks of 2 to 5 m in height occur at 120 to 500 m apart from each other. The length and width of these hummocks are 300 to 2000 m and 250 to 5000 m, respectively. The longitudinal and transverse ridges are 2 to 4 m in height, 250 to 1250 m in length and 120 to 250 m in width and occur 250 to 500 m apart from each other. The slopes of the windward and flanks of the sand ridges are 1° to 2° and 4° to 6°, respectively. The size of sand grains varies from 0.06 to 0.18 mm and more than 49 per cent sand grains are of 0.06 to0.07 mm diameter.

The soils of these dunes are calcareous and non-calcareous fine sand. The moisture within the dunes has been encountered at 30 to 40 cm depth and the percentage of moisture varies from 2.3 to 2.5. These dunes support *Aristida funiculata, Cenchrus biflorus*, and *Crotalaris burhia*. Such dunes are visible in the form of specks on Landsat band 7 black and white image in grey to drak grey tone (Fig. 12.3).

DIGITAL LANDSAT SPECTRAL CHARACTERISTICS OF SAND DUNES

The parabolic dunes in light tone stand out conspicuously in continuous chains on Landsat imagery and occupy 80 per cent of the dune tract. The longitudinal and transverse dunes distinctly appear as linear feature in alternating light to dark grey tone on Landsat images of all four bands and emerge with the greatest clarity on Landsat MSS band 7 image of dry and wet seasons. Variations in the moisture and vegetation pattern are easily recognised on Landsat MSS band 5 image. The flanks and crests of these dunes portray a light tone on this image due to the thick accumulation of fresh sand Shankarnarayan and Singh (1979).

The spectral characteristics of the sand dunes are clearly visible on the computer print out of Landsat MSS band 7 by the dark and light pixels in six grey levels. The number of pixels, grey levels, brightness values and acreage of each classified dune are presented in Table 12.1.

Table 12.1. Number of pixels, brightness values and acreage for the classified sand dunes of a sub-scence in western part of the middle Luni basin (Computer printout of Landsat MSS band 7)

S. No.	Sand dunes	Grey levels	Brightness values	No. of pixels	Area (km^2)	Percentage (%)
1.	Deflation hollows (swales) of sand dunes	1-2	0-26	831	3.75	17.7
2.	Sand dunes with good moisture regime	3	27	1304	5.88	27.8
3.	Sand dunes with poor moisture regime	4-5	28-29	2159	9.63	46.1
4.	Bare sand dunes	6	30-32	396	1.79	8.4

Source : Singh and Shankarnarayan (1986).

The parabolic sand dunes, because of their characteristics shape, size and pattern are distinctly visible on Landsat MSS band 7 computer printout. The windward slope (deflation hollows) of the parabolic sand dunes and sand dunes having better moisture status with brightness values (0-26 & 27) appear in dark pixels, while flanks and crests of the dunes with brightness values (28-29) appear in dark grey to gray pixels. The bare sand dunes, because of high reflectance in infrared radiation and brightness values (30-32) appear in white tone (Singh and Shankarnarayan, 1986). The spectral variations on different aspects of the sand dunes obviously indicate the variations in soil texture, moisture status, vegetation pattern and size, shape and nature of the sediments. It is significant to remark that these variations within the sand dunes could not be detected from the visual interpretation of Landsat images.

MANAGEMENT AND STABILISATION OF SAND DUNES

The sand dunes based on the morphological, pedological and vegetation characteristics and moisture regime have been classified into two dune system viz; old and new (Fig. 12.4) for their proper management and stabilisation.

Management of Sand Dunes of Old System

The obstacle, parabolic, longitudinal and transverse sand

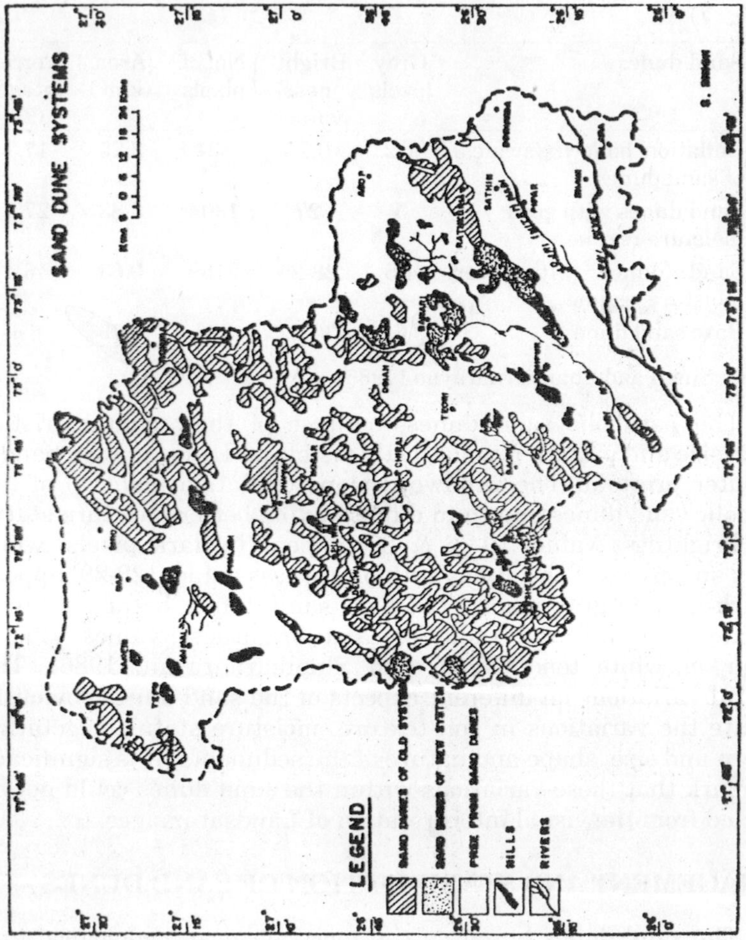

Fig. 12.4. Sand dunes of old and new dune systems

dunes belong to the old dune system. These dunes are well stabilised, fossilised, vegetated and rills, gullies and alternate layers of calcium carbonate layers are developed and are running in eluvial phase.

The morphological, pedological and vegetation characteristics and nutrient and micronutrient status of these dunes indicate that stabilisation measures are not needed to ensure their stability but they need following measures for their management.

(i) Stabilised and fossilised dunes free from fresh sand deposits could be managed by protecting them from the onslaught of man and his animals.

(ii) Reactivated parts of the stabilised dunes covered with fresh sands should be protected from erosion by mulching and due to fast regeneration of vegetation grasses and shrubs will come up in due course of time.

(iii) The cultivation on the dunes of old system should be stopped and they could be developed into silvipastures by suitable grass and tree species.

Stabilisation of Sand Dunes of New System

The active barchan and shrub-coppice dunes, sand sheets and sand ripples fall under the category of new dune system. These dunes and sands are characterised by loose sands, low content of calcium carbonate and absence of vegetation but better status of moisture throughout the profile. The dunes, therefore, are highly susceptible to wind erosion and are posing serious problems to the roads and other engineering structures. These sand dunes could be stabilised by adopting following phyto-reclamation techniques suggested by Bhimaya *et al.,* (1961) which will involve the following three steps.

1. Protection against biotic interference by fencing dunes with 5 strand barbed wire fixed on angle iron posts spaced 3 to 4 m apart.

2. Treatment of active dunes by fixing brushwood barriers on windward side of the dunes in 2 to 5 m apart parallel lines, oriented across the prevailing wind direction and in case of multiple wind direction in the chess board pattern. The plant species such as *Leptadenia pyrotechnica* (Khimp); *Calligonum polygonoides* (Phog) and *Lasiurus sindicus* (Sewan) have been found useful and economical for this work.

3. Sowing of grasses and transplanting of trees and shrub species raised in earthen bricks of 30 cm height with a cross section of 10 cm^2 at top and 15 cm^2 at bottom. These bricks are prepared

from a mixture of sand, clay and F.Y. manure mixed in equal proportion. The plantation bricks have the advantage of providing nutrition and anchorage to seedlinge during their early stages of establishment. During planting season, these bricks with seedlings may be carried to the planting sites and could be planted deep at spacing of 5m x 5m so that the bricks containing seedling roots are placed in the moist sand layers.

13

IMPACT OF PRESENT AND PALAEO DRAINAGE SYSTEMS ON GEO-ENVIRONMENT USING REMOTE SENSING TECHNIQUES

INTRODUCTION

In India, the present and palaeo rivers and their tributaries had played a significant role in the evolution of cultural and biophysical landscape under different climatic zones. The palaeo drainage channels have also provided the ideal sites for the development of different type of cultures and civilisations as evidenced by the presence of the old settlements and objects of domestic utility at the archaeological excavated sites along their buried courses.

In the Indian desert and adjoining regions of north-west India, the present and palaeo Aravalli and Himalayan drainage systems viz; the Luni-Jawai, the Mahi, the Sabarmati, the west Banas, the Banas-Berach, Kothari-Bandi, the Sota-Sahibi, the Dohan-Kantli and the Saraswati-Drishadvati-Sutlej (Satadru) and the Ganga-Yamuna have been identified and mapped using remote sensing and ground survey techniques. These drainage systems had positive and negative impacts on the geoenvironment of the Indian desert and adjoining regions as evidenced by the morphological, morpho-hydrological, morphopedological, morphoecological features and environmental problems.

The positive and negative impacts of the above present and palaeo drainage systems on the geoenvironment of the Indian desert and adjoining regions have been discussed in this chapter.

POSITIVE IMPACT OF PRESENT AND PALAEO DRAINAGE SYSTEMS ON GEOENVIRONMENT

The present and palaeo drainage channels of the Luni-Jawai and the Saraswati-Drishadvati-Sutlej drainage systems (Fig. 13.1) have significantly contributed in the evolution of alluvial plains and saline depressions, ground water development and landuse potentials (Fig. 13.2 & Fig. 13.3).

Evolution of Alluvial Plains

In the Luni-Jawai drainage basin (Fig. 13.2), the Takhatgarh-Mitri, Sukri-Bandi, Kundal-Ramania, Golia-Bhailan, the Khari-Bandi-Sagi, Harji-Menadar-Las, the Khari, Krishaoti and the Jojri-Mitri-Lik are the most significant older alluvial plains. The depth of the alluvium in these plains varies from less than 1 m to more than 30 m depending on the terrain conditions of the basin. The texture of the sediments varies from sand to loamy sand, sandy loam to loam and also silty clay loam. These plains form the backbone of agricultural production (Singh, 1992; Ghose, 1965). The present and palaeo drainage channels of the Saraswati- Drishadvati-Sutlej (Fig. 13.3) drainage system have created the older alluvial plains in the northern and north-eastern parts of the Indian desert. The thickness of the alluvium in these plains ranges between 5 to 200 m and even more at few places. The dominant alluvial plains occur in Hissar-Hansi-Sirsa-Hanumangarh, Churu-Jhunjhunu-Taranagar- Jodhpur, and Ganganagar-Bikaner-Jaisalmer-Barmer tracts. The younger alluvial plains also occur in narrow strips along the major rivers of these drainage systems (Fig. 13.2 & Fig. 13.3).

In the east of the Aravalli mountains, the vast alluvial plains created by the Chambal-Banas and the Ganga-Yamuna drainage systems occur right upto western parts of Uttar Pradesh. The depth of the alluvium in these alluvial plains varies from 10 to 1000 m or more. The younger alluvial plains lying on either side of these rivers are more extensive and have more thick alluvial deposits than that of the desert drainage systems.

Development of Water Resources

The present and palaeo drainage systems also have a direct impact on the development and distribution of surface and ground water. In the desert regions, there is an acute shortage of water but the present and palaeo drainage channels and older and younger alluvial plains have provided the ideal hydrogeomorphic sites for the

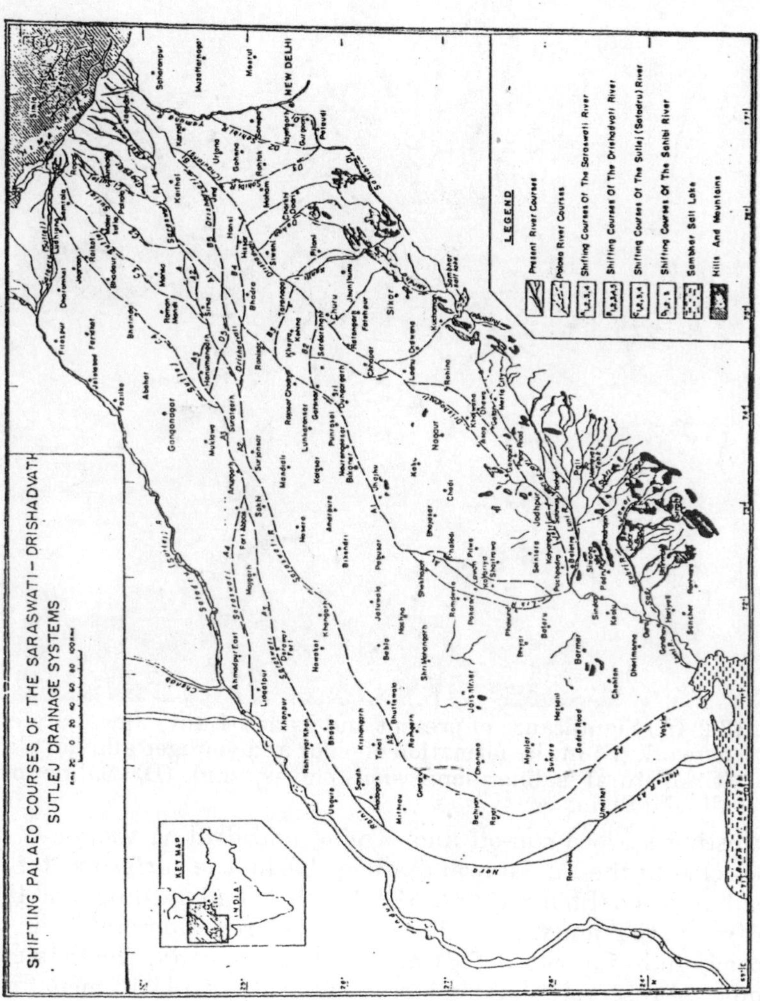

Fig. 13.1. Shifting palaeo courses of the Saraswati - Drishadvati- Sutlej drainage systems

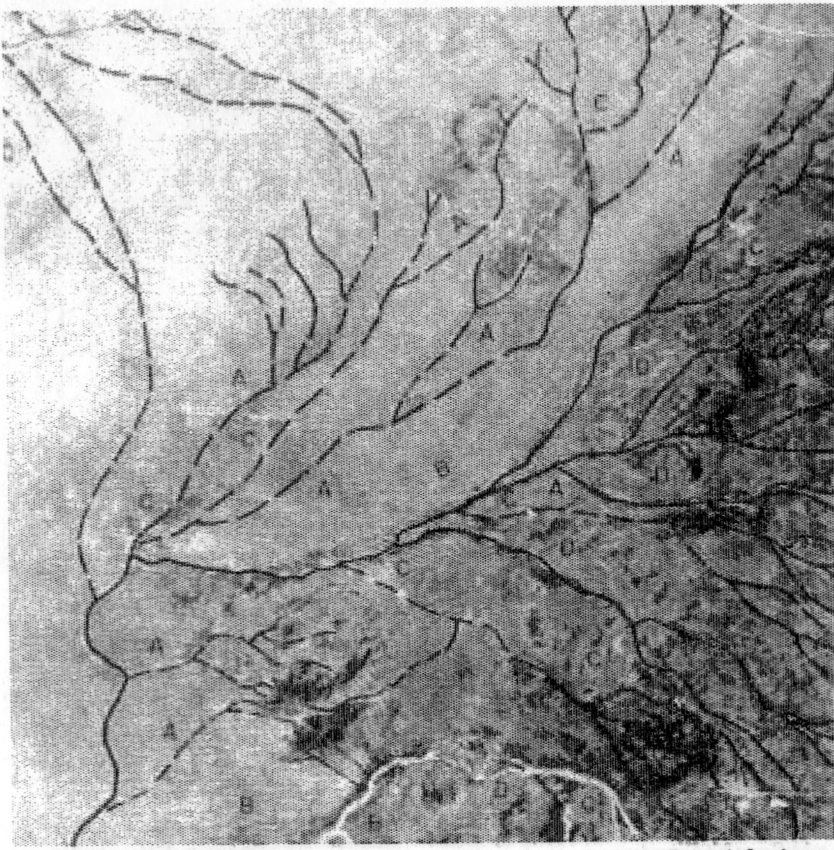

Fig. 13.2. (A) Significance of present and palaeo Luni - Jawai drainage network (B) In the formation of older and younger alluvial plains (C) Natural saline depressions/playas, and (D) Saline older alluvial Plains

harvesting surface run-off and digging and drilling the wells and tubewells. In the Luni-Jawai drainage basin, the surface water has been harvested (Sharma *et al.*, 1989) in the small, medium and large size reservoirs and tanks. There are 138 reservoirs/tanks in the upper Luni basin with 356.81 mcm storage capacity. The maximum number of reservoirs having storage capacity of 141.16 mcm are in upper Luni catchment, followed by the Guhiya, the Bandi and the Jojri catchments having 17, 16 and1 reservoirs with storage capacity of 127.19, 88.10 and 0.36 mcm, respectively. Out of the total reservoirs, Sardar Samand reservoir constructed at the confluence of the Sukri and the Guhiya rivers have the largest water storage

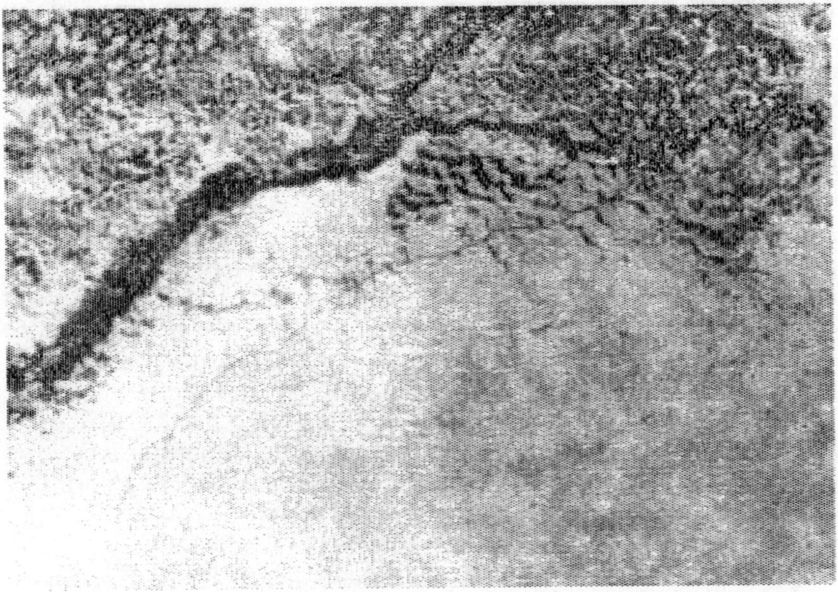

Fig. 13.3. Significance of present and palaeo Saraswati-Drishadvati drainage network in the formation of saline and non-saline older and younger alluvial plains, natural saline depressions/playas and development of aquifers

capacity (88.24 mcm); followed by Hemawas tank along the Bandi river (62.55 mcm); Jaswantsagar along the Luni river (52.84 mcm); Surpura tank (21.65 mcm); Kharda tank on the Phunpharia stream (8 mcm) and the Chaupra tank (Fig. 13.2) on the Reria stream (7 mcm) (Shankarnarayan and Kar, 1983). In the east of the Aravallis, several reservoirs/tanks have been constructed along the rivers of the Chambal-Banas and the Ganga-Yamuna drainage systems for irrigation and drinking water purpose. The anicuts across the existing minor tributaries have also been constructed for water harvesting and recharge of ground water.

The older and younger alluvial plains created by well integrated Luni-Jawai and Saraswati-Drishadvati-Sutlej drainage systems during the late Quaternary period have provided ideal hydrogeomorphic sites for the development of older and younger alluvial aquifers at varying depth. In these aquifers, the inter-granular porosity of the unconsolidated sediments determines their permeability. In the older alluvial aquifers, the depth to water table varies from 10 to 100 m below ground level (b.g.l) and the yield of wells ranges from 10 to 70 cum/hr. Whereas in the younger alluvial aquifers, the depth to water table varies from 8 to 30 m b.g.l. and yield of water from wells and tubewells is 20 to 50 cum/hr. The depth to water table and yield of the older and younger alluvial aquifers developed under alluvial plains of the Himalayan and the Aravalli drainage systems (Fig. 13.2 & Fig. 13.3) are variable depending upon the nature and thickness of alluvium ranging from 6 to 25 m and 100 to 800 cum/hr (Chatterji *et al.,* 1979).

The buried courses of the palaeo stream channels of the Luni-Jawai and the Saraswati-Drishadvati-Sutlej drainage systems have provided the ideal geomorphic sites for the development of shallow to moderately deep aquifers (Fig. 13.2 & Fig. 13.3). The buried wide courses filled with sand and gravels regulate the flow of water under different geomorphic settings and the depth to water table varies from 5 to 20 m and yield of water from wells is 5 to 30 cmm/hr (Singh, 1976-77; Ghose, 1965). It has been qualitatively estimated that the buried course of the Saraswati to the west of Jaisalmer has a reserve of about 3000 mcm of ground water (Venkateswarlu *et al.,* 1990). The palaeo buried courses of the Saraswati-Drishadvati-Sutlej still get water from the eastern higher rainfall Himalayan catchment and they facilitate subterraneous movement of water from higher rainfall areas to the arid and semi-arid areas.

Origin of Saline Depressions and Salts

In the Indian desert, saline depressions of different shape and size such as Pachpadra, Sanwarla, Thob, Lunkaransar, Kuchaman, Talchhapar, Didwana and Sambhar occur near the confluences and also along the courses of the palaeo buried streams (Fig. 13.2 & Fig. 13.3). These saline depressions and their salts have originated due to the fluvial and aeolian proceses and capillary rise and evaporation of salt brine under two major wet and two major dry phases during late Quaternary period. The major salts and evaporites of commercial value extracted from these saline depressions are sodium chloride and calcium sulphate which have provided employment and improved socio-economic conditions of the local people.

Cultural and Biogeographical Environment

Present and palaeo drainage systems have also influenced the cultural and biogeographical environment such as the roads, railway tracks, settlements, landuse and vegetation. The levees and other higher parts of the old and younger flood plains have been used for the construction of roads, railway tracks and settlements as they are considered to the safe sites for the unforeseen events.

The younger and older alluvial plains, filled valleys and old flood plains of the Luni-Jawai and the Saraswati-Drishadvati- Sutlej drainage systems have very good agricultural potentials for growing rainfed and irrigated crops due to fertile soils and better ground water potentials. The old and younger alluvial plains occurring along the palaeo and present drainage channels are used for double cropping which provide higher crop yield (Singh, 1992). The older alluvial plains, by and large, are put under rainfed crops. The alluvial plains form the backbone of the agricultural production. The growth and distribution of halophyte and phreatophyte plants are also controlled by the present and palaeo drainage channels. The palaeo and existing river courses and their flood plains are supporting various type of plant species viz; *Haloxylon salicornicum, Sporobolus marginatus, Prosopis juliflora, Salvadora persica, Atriplex, Prosopis cineraria, Acacia nilotica* ssp. *indica, Acacia arabica* and *Salvadora oleoides*. The human and livestock populations are largely concentrated in the above alluvial plains (Fig. 13.2 & 13.3).

Palaeoclimate and Ancient Civilisation

The stratigraphic sequence of the sediments along the buried courses of palaeo stream channels and saline depressions has provided the evidences of the past climatic fluctuations under wet and dry phases. The archaeological excavations at Kalibangan located on the Ghaggar river near Pilibangan and Suratgarh and at Kunal and Banawali villages located on the Saraswati river in Hissar district have provided the information about settlements patterns, articles of domestic utility and religious practices. These archaeological evidences clearly indicate that the civilisation of Pre-Harappan and Harappan period flourished around 5 to 4 ka BP during the second major wet phase which prevailed from 10 - 3.8 ka BP. With the establishment of secondary phase around 3.8 ka BP, the Saraswati (Ghaggar) and the Drishadvati rivers were desiccated, disorganised and disappeared and Kalibangan and other towns and villages located on these rivers were also abandoned and the population of these settlements was migrated to other safer areas.

NEGATIVE IMPACT OF PRESENT AND PALAEO DRAINAGE SYSTEMS ON GEOENVIRONMENT

The increasing biotic impact on the present and palaeo drainage systems where man is acting as an important geomorphic agent has resulted in the devastation of agricultural lands, engineering structures, human settlements and livestock.

Impact on Agricultural and Grazing Lands

The salinity/alkalinity, flood and pollution hazards developed due to the human interference with the present and palaeo drainage systems have devasted and depleted the biological productivity of the agricultural and grazing lands. Due to the construction of the reservoirs/tanks, canals and anicuts along the present and palaeo buried stream channels, the salinity/alkalinity and waterlogging conditions have developed resulting in the degradation of geoenvironment and depletion of the crop yield and biomass (Ghose *et al.,* 1982 & Fig. 13.4). The unprecedented occasional floods, water-pollution and accelerated water erosion in the form of rills and gullies have also affected and decreased the yield of crops and biomass from 10 to 15 q/ha to 2 to 4 q/ha and 2 to 4 t/ha to 0.5 to 1 t/ha, respectively. The courses of the Luni-Jawai rivers have widened by 1.8 times (Sharma *et al.,* 1989) during the period of 28 years (1958-1986) and the Kosi river course has shifted to 120 km towards west in last 250 years resulting in the dissections of the flood plains and devastation of agricultural lands.

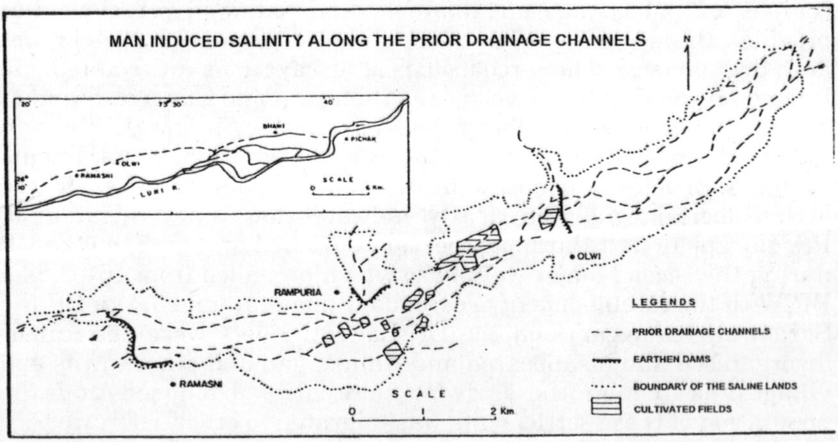

Fig. 13.4. Man induced salinity along the prior drainage channels

Impact on Engineering Structures

In these drainage basins, three major geomorphic flood zones viz; younger alluvial plains and river beds, old flood plains and buried courses of palaeo stream channels and hills and pediment zones are highly unstable and very frequently liable to flood hazards.

The engineering structures like dams, railway tracks, bridges, culverts, roads and irrigation network constructed in these geomorphic flood zones were washed away during the floods of 1979, 1982 and 1990. The interactions among man, land and water are largely responsible for such negative impact on the above engineering structures. Jaswantsagar, Berai and several earthen dams due to their unfavourable geomorphic locations were also washed away (Singh, 1992).

Impact on Life and Property

Several human settlements located on the present and palaeo buried courses of the drainage channels of the Luni-Jawai and the Saraswati-Drishadvati drainage systems were devasted and several animals and human beings were died during the flash floods of 1979 to 1990. More than 50,000 houses located in the geomorphologically unstable areas were either completely destroyed or seriously damaged resulting in a total loss of about Rs. 0.20 million.

14

REMOTE SENSING IN MONITORING AND COMBATING DESERTIFICATION

INTRODUCTION

Desertification is one of the serious environmental and socio-economic global problems that has attracted the attention of the environmentalists, planners, policy makers, politicians, scientists, common people and non-government organisation (NGOs). This serious environmental problem occurs in 5.2 billion hectares of dryland of the world excluding the hyper-arid areas and affects 3.6 billion hectares which is about 70 per cent of the world ms drylands and one quarter of the total area of the world. In these drylands, 990 million people residing in 100 countries of the world are at risk due to the degradation and depletion of their biological productivity.

STATUS OF DESERTIFICATION

The definitions of desertification according to the specialists of various subjects are legion. But according to the latest definition approved by 1992 UN Conference on Environment and Development (UNCED) held at Rio de Janeiro in Brazil, desertification is defined as *land degradation in arid, semi- arid and dry sub humid areas resulting from various factors, including climatic variations and human activities*.

Based on the above definition, the present desertification status map of western Rajasthan (Fig. 14.1) was prepared using multidate satellite imagery and existing natural resource maps and reports (Singh *et al.*, 1992). The present desertification status map of western Rajasthan involves four severity classes of desertification viz., slight, moderate, severe and very severe and their impact on four major landuse systems such as rainfed croplands, irrigated

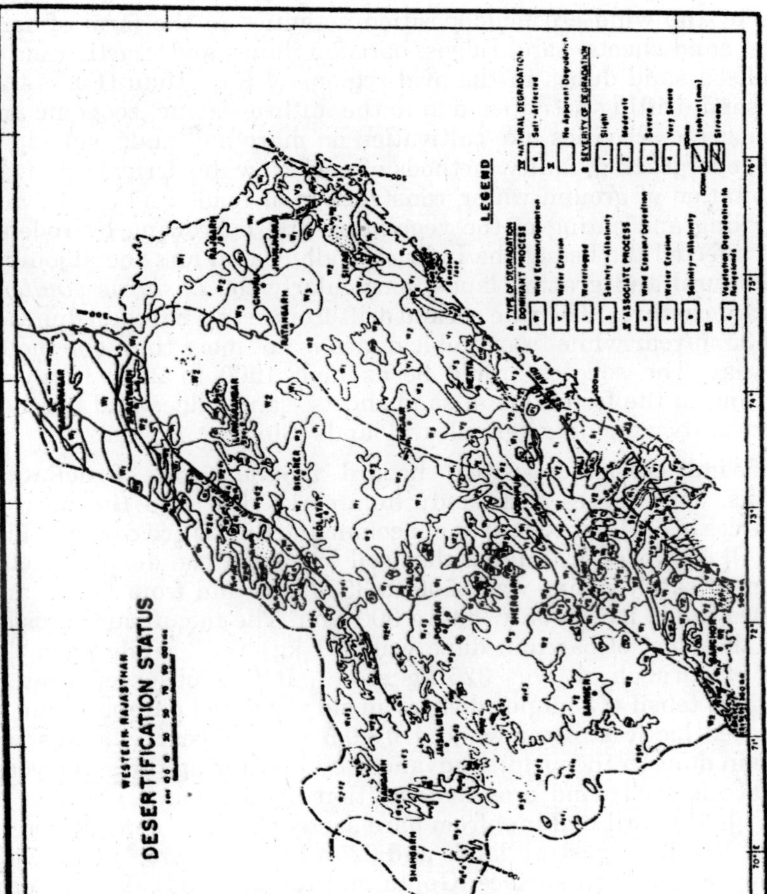

Fig. 14.1. Desertification status

croplands, rangelands (grazing lands) and forests. The desertification processes responsible for the degradation and depletion of the productivity of the above four major landuse systems are wind erosion/deposition, salinity/alkalinity, water erosion and vegetation degradation.

The greatest menace to the agricultural lands., grazing lands, roads, railway tracks, canals, lakes and other water bodies is being done by the wind erosion/deposition activities in the form of sand drifts, sand sheets, sand ridges, barchan dunes and reactivation of stabilised sand dunes in the arid regions of Rajasthan (Fig. 14.2). The sand drifting activities due to the anthropogenic, zoogenic and technogenic activities like cultivation on marginal lands, cutting of trees, overgrazing, faulty methods of surface water irrigation, over-exploitation of ground water, construction of roads and canals and trampling and eating of the vegetation and burrowing by rodents have affected the bed of the Indira Gandhi Canal and the adjoining agricultural and grazing lands, particularly the *Lasiurus sindicus* (Sewan) pastures. The rate of sand drift from active barchan dunes is upto 35 m/year while from stable dunes is not more than one meter per year. The soil loss could be as high 1500 to 2500 t/ha/year resulting in the formation of sand sheets, sand ridges and barchan dunes on the adjoining agricultural lands (Singh *et al.,* 1992).

Wind erosion/deposition hazard in the form of deflation hollows, sand dunes and sandy hummocks degraded the natural resources of 68.4 per cent area of aeolian and fluvial ecosystems (Fig. 14.1). It has been estimated that soil loss from the deeply tractor ploughed sandy plains was 2500 to 3500 t/ha and from fallow and untilled lands it was only 200 to 300 t/ha. The mean sand erosion from the crest of barchan dune was 351 kg $m^{-2}d^{-1}$ while from the crest of parabolic it was 325 kg $m^{-2}d^{-1}$. It was observed that in Shergarh tehsil of Jodhpur district an old vegetated parabolic due to increased biotic stress advanced by 0.5 m in a year whereas the barchan dune in the vicinity advanced by 7 m during the same year. The accelerated wind erosion/deposition resulted in a soil loss of 1449 t ha^{-1} (April to June) from bare sandy plains in comparison to no soil loss from pasture lands and 225 t ha^{-1} from crop lands with crop residue on soil surface (Gupta and Raina, 1994). The general trend of sand movement in this zone is 2 to 20 m/year. As a result, the biological productivity under different type of landuse systems has been depleting.

Water erosion, floods, mining and industrial effluent degraded 11.2 per cent area of the fluvial and aeolian ecosystems in the form of rills, gullies, dissections, siltation and surface crust (Fig. 14.1). In the Luni basin, the accelerated water erosion removed 0.2 to 53.5 gl^{-1} sediments in the hilly terrain of Aravallis and 20.5 to 453.6 gl^{-1} in the lower reaches of the basin. The rills and gullies of 0.5 to 10 m depth and 1.2 to 15.3 m width created by fluvial action degraded the hills, rocky/gravelly and buried pediments and flat older alluvial plains (Fig. 14.2). It has also been estimated that over a period of 28 years (1958-1986) gullied area increased from 199 ha to 242 ha. It has been estimated that during the flash floods of 1990, large quantity of sediments were eroded and deposited in the form of sand sheets of 20 to 60 cm thickness and sand mounds and sand ridges of 0.7 to 2.5 m height, resulting in the depletion of biological productivity of original sediments.

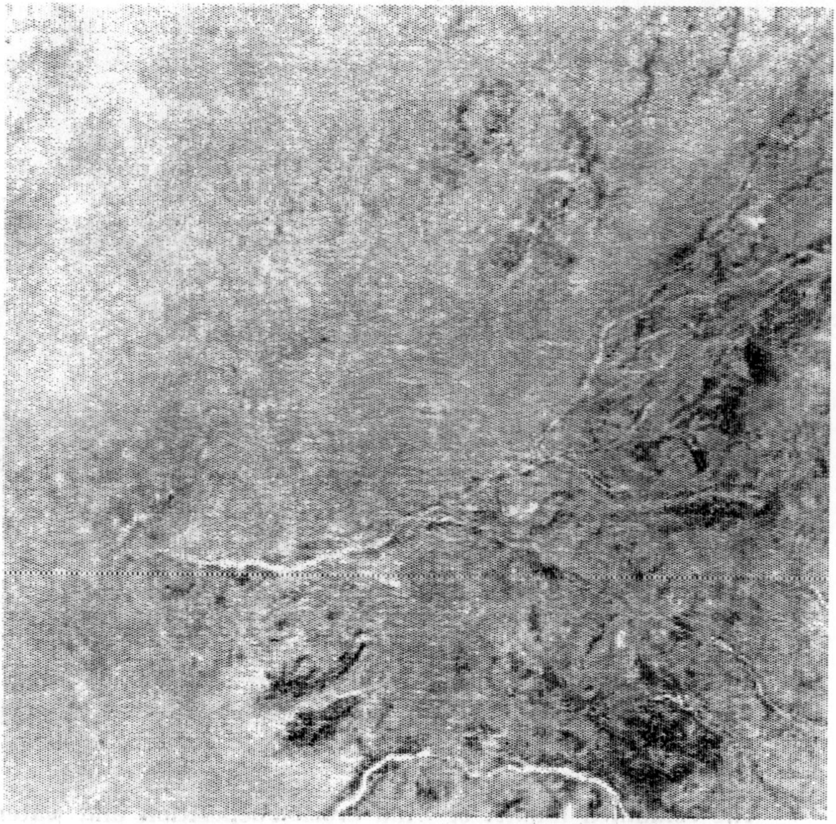

Fig. 14.2. Landsat band 5 black and white imagery of the Luni-Jawai drainage basin showing the tonal and spatial characteristics of the areas desertified by wind erosion/ deposition, water erosion and salinity alkalinity

Industrial effluent along the Jojri, the Bandi and the Luni rivers in Jodhpur, Pali and Barmer districts has affected 34,500 ha area of river bed, younger and older alluvial plains (Fig. 14.3). Due to this hazard the ground water has been very severely, moderately and slightly polluted (EC 2.5 - 45.6 dSm^{-1}). The irrigation by this polluted water has changed the morphological and physico-chemical properties of the soils, resulting in the reduction in their infiltration rate and fertility status. The biomass, density and cover of the plants in the affected areas also declined by 3 to 5 times. The cropping intensity, crop yield and crop residue declined by 60 to 70, 40 to 50 and 30 to 40 per cent, respectively.

Fig. 14.3. IRS-LISS II FCC imagery showing the areas affected by industrial effluent along the Luni river near Balotra and Jalor

It has been worked out that out of total area along these rivers, 19.39, 24.48, 33.73 and 22.40 per cent area are affected by very severe, severe, moderate and slight category of severity of degradation, respectively.

The combined impact of wind erosion/deposition and water erosion affected the natural resources of 7.3 per cent area of sandy undulating buried pediments and obstacle dunes (Fig. 14.1). Very severe wind erosion/deposition and severe water erosion created the sand dunes and hummocks of 5 to 30 m height and gullies of 6 to 15m depth and 8 to 25 m width, resulting in the degradation of large acreage of the above ecosystems (Fig. 14.2). Moderate to severe wind erosion/deposition and moderate water erosion hazards were responsible for creating sand dunes of 3 to 12 m height, sand sheets of 250 to 450 cm thickness and gullies of 2.5 to 5.6 m depth and 5.5 to 10.5m width. Slight to moderate wind erosion/deposition and water erosion hazards created sand dunes and hummocks of 1.5 to 5 m height and gullies of 0.5 to 3.5 m depth and 2.5 to 4.5 m width.

Salinity/alkalinity and waterlogging problems (Fig. 14.1) developed in Indira Gandhi Canal Command area and Luni-Jawai drainage system due to faulty methods of irrigation, presence of subsurface barrier at shallow depth and construction of tanks and reservoirs along the courses of palaeo drainage channels have affected the natural resources of 3.1 per cent area of different fluvial and aeolian ecosystems (Singh *et al.,* 1992). Recent field investigations revealed that in Suratgarh tehsil of Ganganagar district and Rawatsar tehsil of Hanumangarh district 4,500 and 5,000 ha agricultural lands, respectively have been desertified by waterlogging and salinity/alkalinity problem. Due to this problem, natural cover of *Haloxylon salicornicum* and *Lasiurus sindicus* is being depleted at a faster rate. In the catchments of the Sardar Samand and Jaswant Sagar reservoirs, 9600 and 6170 ha area, respectively have been degraded due to seepage and suspended salts and productivity of large acreage of agricultural lands has been depleted (Fig. 14.2).

Natural desertification hazard occurring in the west of 100 mm rainfall zone affected in natural resources of 2.1 per cent area of the aeolian ecosystem. Remaining 8.1 per cent area is free from such hazards (Fig. 14.1).

As a result of different human-induced desertification hazards, the vegetation cover, particularly in rangelands, has been depleting. The basal cover of *Lasiurus sindicus* and *Eleusine compressa* grasses declined from 5.14 to 2.40 per cent. The total vegetation cover declined from 6.59 to 0.95 per cent and its yield declined by half (Fig. 14.1). The rocky/gravelly pediments due to increasing degradation support very poor cover of *Dactyloctenium sindicum-Eleusine* type of grass (Fig. 14.2). The basal cover of these plant species declined from 2 to 4 per cent to less than 1 per cent and yield changed from around 1 t/ha to 0.27-0.45 t/ha under protected and unprotected areas, respectively.The density of *Salvadora* declined from 10-20 trees/ha to 1-3 trees/ha (Shankar and Kumar, 1987).

Due to increased human activities and accelerated erosion, the vegetation cover of *Prospis cineraria, Capparis decidua, Ziziphus nummularia, Cynodon dactylon, Cenchrus ciliaris* and *Dicanthium annulatum* decreased from 4-8 to 1-2 per cent and biomass yield from 4000 kg/ha to 130-1100 kg/ha (Saxena, 1985).

DIGITAL ANALYSIS OF LANDSAT DATA FOR ASSESSMENT OF LAND VULNERABILITY TO DESERTIFICATION

Digital analysis of Landsat band 5 computer print out comprising the western part of the central Luni basin revealed various degree of vulnerability of the aeolion landforms to desertification process (Shankarnarayan and Singh, 1979).

The extent of the vulnerability of areas to desertification has been computed and presented in Table 14.1. The landforms with dark pixels due to the presence of vegetation and moisture (brightness value 0-65) occupy 3.36 km^2 area of the Landsat subscene are least vulnerable to desertification. Landforms with light to dark grey pixels due to moderately sparse to sparse vegetation (brightness value 66-69) constituting 6.07 km^2 area are moderately vulnerable to desertification. Landforms with bright pixels due to the presence of fresh sand and absence of vegetation (brightness vlaue 70-83) occupy 11.72 km^2 area are highly vulnerable to desertification.

Table 14.1. Vulnerability assessment of the aeolion landforms to desertification

Degree of vulnerability	Grey level	Brightness value	Number of pixels	Extent (km^2)	Percentage (%)
Least	1	0-65	745	3.36	15.9
Moderate	2-3	66-69	1346	6.07	28.8
High	4-6	70-80	2599	11.72	55.3
			4690	21.15	100.00

MEASURES FOR COMBATING DESERTIFICATION

It is apparent from foregoing discussions that high human population density and growth rate, cultivation on marginal lands, deep tillage and injudicious methods of irrigation are responsible for desertification of natural resources of different fragile arid ecosystems. The following control/management/rehabilitation measures for combating desertification of the areas which are not yet degraded or likely to be degraded or slightly degraded by involving government and non-government organisations, local people including men and women, international and regional organisations are recommended.

Control of Wind Erosion/Deposition

Development of large aggregates or clods to resist wind action, creation of roughness on land surface to reduce wind velocity, establishment of barriers or traps (wind breaks/shelterbelts) to reduce wind velocity, stabilization of sand dunes, creation of vegetative or plant residue cover on soil surface (stubble mulching), minimum tillage and conservation of rain water are some of the measures found effective in checking wind erosion (Gupta and Raina, 1994). Crop residues of 2 to 5 t/ha and pearl millet stubbles of 45 cm height were found very effective in preventing the blowing of sand from sandy soils. Strip cropping of erosion susceptible and erosion resistant perennial grasses reduced the impact and threshold velocity of wind to the minimum and thus checked the erosion and increased the production of crops grown in between the protective strips.

Sand dune stabilization is very important technique in checking wind erosion. Though, there are different methods of sand dune stabilization, this can best be done by vegetation. Before raising vegetation, the sand dunes are fixed either chemically or with mulches using locally available dead grasses or brush wood, etc. Usually checker board or parallel strip methods are followed. After fixation, the sand dunes are afforested with fast growing species of trees and grasses. *Acacia tortilis, A. radiana, Prosopis juliflora, P. cineraria, Ziziphus mauritiana* trees, *Calligonum polygonoides, Crotalaria burhia, Aerva javanica, Z. nummularia* shrubs and *Lasiurus sindicus, Panicum turgidum* and *P. antidolate* grasses have been reported to be most useful species for the afforestation and stabilisation of sand dunes in the arid areas of western Rajasthan, India (Kaul, 1985).

Control of Water Erosion

The problem of water erosion control should be taken up on watershed basis rather than in a piecemeal. The construction of check dams, contour bunds, contour furrows, graded bunding, bench terracing and contour trenching could be used to check water erosion under different landuse systems. Vegetative barriers of perennial grasses like *Cenchrus ciliaris, Cenchrus setigerus* and *Cymbopogon jwarancusa* on contour lines have been proved successful in conserving moisture and increasing crop production. The run-off volume reduced between 2 to 71 per cent, conserved soil moisture increased from 2 to 13 per cent and yield of clusterbean increased between 37 to 51 per cent in western Rajasthan, India (Faroda, 1996). These barriers are easy to raise, less expensive and provide fodder during lean period.

Management of Saline/Alkaline and Waterlogged Areas

With the efficient management of surface and ground waters, the saline/alkaline and waterlogged areas can be managed. In the areas having impervious strata at shallow depth, crops requiring less water should be grown. Sprinkler and drip irrigation methods should be adopted. In saline areas after salt leaching, salt tolerant/salt loving crops/varieties should be grown. The areas affected by alkalinity could be managed by applying gypsum at the rate of 100 per cent gypsum requirement, followed by leaching (Joshi and Dhir, 1991). The waterlogged areas could be managed by providing vertical drainage, bio-drainage and surface and subsurface drainage. After assessing the quality of water, the stagnated water could be transferred to canal and rainfed areas for irrigation and reduction of canal water allowance from 5.23 to less than 3 cusec/1000 acres.

Rehabilitation of Mining Areas

Mining areas which are left uncared can be revegetated with proper soil and water management practices. For gypsum mined areas (Kavas) micro-catchment and half-moon terraces for rainwater harvesting, profile modification with farm soil and FYM and planting of *Tamarix aphylla, Pithecellobium dulce, Salvadora oleoides* and *Cercidium floridum* have been found to be successful in rehabilitating them. In case of limestone mined areas (Barna), inward sloping bench terraces, soil profile modification by addition of farm soil, bentonite and FYM, and planting of *Acacia catechu, A. tortilis, A. nubica, A. senegal, A. planifrons* and *Cercidium floridum* have been found to be successful in rehabilitating these wastelands (CAZRI, 1995-96).

Management of Industrial Effluent Affected Areas

In order to minimise the impact of industrial effluent water and desertification of natural resources, this water should be properly treated at source point and should be safely drained out in the river bed. The concentration of pollutants may be reduced by adding fresh water before discharging them into tanks. Combination of fly ash and gypsum should be used for removing the colour and reducing the pH of the effluent water.

In the course textured soils (sand to loamy sand), the gypsum at the rate of 3 to 11 t/ha should be applied. Whereas in the medium textured (sandy loam) soils 6 to 15 t/ha and in fine textured (loam to clay loam) soils 10 to 23 t/ha gypsum should be applied.

The trees viz., *Eucalyptus camaldulensis, Acacia nilotica, Acacia tortilis, Prosopis cineraria, Prosopis juliflora* and *Tecomella undulata* and salt tolerant grasses such as *Chloris virgata, Sporobolus helvolus, S. marginatus* and *Dicanthium annulatum* should be grown in industrial effluent (Singh and Balak Ram, 1997).

REFERENCES

Abichandani, C.T. 1965. Use of aerial photographs in soil survey of arid zone of western Rajasthan. *Ann. Arid zone* 6 : 220-227.

Aggarwal, J.P. and Jadhav, R.N. 1985. Methodology for lTERRA m data analysis and comparative study of aerial, Landsat and TERRA data for forest mapping. *Proc. Six Asian Conference on Remote Sensing, Hyderabad*, pp. 519-523.

Anderson, J.R., Hardy, E.E., Rouch, T.T. and Witmer, R.E. 1976. A landuse and land cover classification system for use with remote senser data. *Geological Survey Professional Paper 964.* U.S. Geological Survey, Reston, Virginia, 28 p.

Ask, R.E. 1952. Elements of photogrammetric optics. In : *Marvel of Photogrammetry* (Ed. American Society of Photogrammetry), George Banta Publishing Co., Menasha, Wisconsim, pp. 17-67.

Basak, P., Prasad, Narasimha, Nazimuddin, M. and Roy, A.K. 1986. Regional ground water resources development in the Chaliyar river basin through remote sensing. *Proc. Int. Semn. Photogramm. and Remote Sensing for the Developing Countries, New Delhi,* Vol.1,pp.T. 4-P/11.1-11.5.

Bhattacharya, A. 1980. Satellite remote sensing in geomorphological survey of Tripura State. *Proc. Semn. Application of Photo-Interpretation and Remote Sensing Techniques for Natural Resources Survey and Environmental Analysis, Dehradun,* pp. 100-104.

Bhimaya, C.P., Kaul, R.N. and Ganguli, B.N. 1961. Sand dune rehabilitation in western Rajasthan. *Sci. Cult.* 27 (5) : 224-229.

CAZRI, 1995-96. Rehabilitation of Jypsum and Limestone areas. *Annual Report,* CAZRI, Jodhpur, pp. 19.

Chari, K.N., Nayak, B. and Mahapatra, G. 1980. Manual interpretation of Landsat-1 imagery for location and estimation of area of the water bodies in Hirakund Comman area, Orissa. *Proc. Semn. Application of Photo-Interpretation and Remote Sensing Techniques for Natural Resources Survey and Environment Analysis, Dehradun,* pp. 182-183.

Chatterji, P.C., Singh, S. and Qureshi, Z.H. 1979. Hydrogeomorphology of central Luni Basin, western Rajasthan (India). *Geoform* 9(3) : 211-224.

Deutsch, M. and Ruggles. Jr.F.H. 1978. Hydrological applications of Landsat imagery used in the study of the 1973 Indus river flood, Pakistan. *Water Resources Bulletin* 14 (2) : 261-274.

Dhir, R.P. 1974. An approach to use of aerial photographs in small scale mapping based on experience of Jodhpur district, Rajasthan. *Journ. Ind. Society Remote Sensing* 2 (1) : 13-18.

Drayton, R.S. 1986. An appraisal of the visual interpretation of standard satellite imagery for surface water planning in the Third World. *Proc. Int. Semn. Photogramm. and Remote Sensing for the Developing Countries, New Delhi,* Vol.1, pp. T.4-P/18.1-18.7.

Dwivedi, R.S. 1985. Evaluation of Thematic Mapper data for soil resources mapping. *Proc. Six Asian Conference on Remote Sensing, Hyderabad,* pp. 418-422.

France, M.J., Collins, W.G. and Hedges, P.D. 1986. A hydrological comparison of Landsat MSS and Landsat TM imagery. *Proc. Int. Semn. Photogramm. and Remote Sensing for the Developing Countries, New Delhi,* Vol. 2, pp. T.9-P/1.1-1.8.

Faroda, A.S. 1996. Indian Thar Desert : Retrospect and Prospect. *Paper Presented at Fifth International Conference on Desert Development, Lubbock*, Texas, USA.

Gautam, N.C. and Narayan, L.R.A. 1988. *Wastelands in India* Pink Publishing House, Mathura, 96p.

Gelmet, R.H. and Gardner, J.V. 1979. Use of radar for water exploration in Nigeria, West Africa. *Proc. Semn. Remote Sensing Applications and Technology Transfer for International Development, Ann. Arbor, Michigan,* U.S.A., pp.42.

Ghose, B. 1965. The genesis of the desert plains in the central Luni basin of western Rajasthan. *J. Ind. Soc. Soil Sci.* 13(2) : 123-126.

Ghose, B., Kar, Amal and Qureshi, Z.H. 1979. The lost course of Saraswati river in the Great Indian desert : New evidences from Landsat imagery. *Geographical Journal* 145 (3) : 446-451.

Ghose, B., Pandey, S. and Singh, S. 1969. Quantitative geomorphology of the drainage basins in semi-arid environment. *Ann. Arid Zone* 8(1) : 37-44.

Ghose, B., Pandey, S., and Singh, S. and Gheesa Lal, 1967. Quantitative geomorphology of drainage basins in central Luni basin, western Rajastha. *Zeitschrift Fur Geomorphologie* 11 (2) : 146-160.

Ghose, B. and Singh, S. 1965. Geomorphological studies in Rajasthan 1. Analysis of landforms of Saila block from aerial photographs. *Ann. Arid Zone* 4 (2) : 207-216.

Gupta, A.K. and Rao, V.R. 1985. Digitally enhanced Landsat imagery for landuse features - A case study for Sirohi district (Rajasthan), India. *Proc. Six Asian Conference on Remote Sensing, Hyderabad,* pp. 428-433.

Gupta, J.P. and Rania, P. 1994. Wind erosion and its control in hot areas of Rajasthan, India. *Proc. Int. Symp. on Wind Erosion in West Africa : The Problem and its Control.* (Eds. B.Burkert, B.E. Allison and M. Von Oppen), University of Hohenhim Germany, pp. 209-218.

Hadley, R.F. and Schumm, S.A. 1961. *Sediment Sources and Drainage Basin Characteristics in Upper Cheyinve River Basin, Washington, D.C.* United States Geological Survey, pp. 137-196 (Water Supply Paper 193 (B).

Joshi, D.C. and Dhir, R.P. 1991. Reclamation of saline sediments deposited during flash flood on agricultural lands in the Indian Arid Zone. *Arid Soil Research and Rehabilitation*, Vol. 5, pp. 175-185.

Kachhwaha, T.S. 1985. Temporal monitoring of forest land for changes detection and forest cover mapping through satellite remote sensing techniques. *Proc. Six Asian Conference on Remote Sensing, Hyderabad*, pp. 276-281.

Kar, Amal, Singh, S., Vangani, N.S., Dubey, J.C. and Ghose, T.K. 1989. Application potentials of IRS-IA data for mapping hydrogeomorphic features in the Indian desert. *Remote Sensing Satellite Mission and its Application Potential, Jodhpur*, pp. 54-58.

Kaul, R.N. 1985. Afforestation of sand dunes. In : *Sand Dune Stabilisation, Shelterbelts and Afforestation in Dry Zone.* FAO Rome, pp. 75-85.

Kolarkar, A.S., Dhir, R.P. and Singh, N. 1980. Salt affected lands in south-eastern arid Rajasthan-their distribution and genesis as studied from aerial photographs. *Proc. Semn. Application of Photo-Interpretation and Remote Sensing Techniques for Natural Resources Survey and Environmental Analysis, Dehradun*, pp. 132-136.

Krishnan, A., Bhati, P.N. and Rakhecha, P. 1966. A soil moisture regime and microclimatological study over dunes in western Rajasthan. *Ann. Arid Zone* 5(1) : 1-9.

Lillesand, T.M. and Kiefer, R.W. 1979. *Remote Sensing and Image Interpretation*. John Wiley and Sons., New York.

Moore, D.G., Michael, E.W. and Myers, Victor. I. 1974. *A Guide for Optical Processing and Use of ERTS-1 Data for Analysis of Surface Water : A Practical Approach*. SDSU-RSI-73-12, U.S.A. 21 p.

Moore, G.K. 1977. Prospecting for ground water with Landsat images. *Proc. U.N. World Water Conference Mar Del Plata, Argentina.*

Murthy, K.N.K., Sharma, K.D. and Vangani, N.S. 1980. *Surface Water Resources of Nagaur District.* Sectional Technical Report, CAZRI, Jodhpur, 47p (mimeog).

Murthy, R.S., Deshmukh, S.N., Raj Kumar and Saxena, R.K. 1980. Soil physiographic relationship studies by using aerial photographs in Chhatarpur district, Bundelkhand region (Madhya Pradesh). *Proc. Semn. Application of Photo-Interpretation and Remote Sensing Techniques for Natural Resources Survey and Environment Analysis, Dehradun*, pp. 169-173.

Murthy, R.S., Sita Ram and Pandit, A.S. 1980. Importance of aerial photo-interpretation in geomorphological mapping. A case study of Rajkot district. *Proc. Semn. Application of Photo- Interpretation and Remote Techniques for Natural Resources Survey and Environmental Analysis, Dehradun*, pp. 105-109.

NRSA, 1979. *Satellite Remote Sensing Survey of Haryana*. Project Report, Vol. 1, 84p.

NRSA, 1979. *Satellite Remote Sensing of Natural Resources of Mizoram*. Project Report, Vol. 1, 133p.

NRSA, 1983. *Nationwide Mapping of Forest and Non-Forest Areas Using Landsat False Colour Composites for the Periods 1972-75 and 1980-82*. Project Report, Vol. 1, Deptt. of Space, Govt. of India, Hyderabad (mimeog).

NRSA, 1983. *Status of Natural Resources Mapping Through Satellite Remote Sensing Techniques*. NRSA, Hyderabad.

NRSA, 1985. *Mapping of Wastelands in India from Satellite Imagery 1980-82*. Deptt. of Space, Govt. of India, Hyderabad.

NRSA, 1985, *Remote Sensing Applications - A Status Appraisal*. Prepared for US-India Symposium Cum Workshop at Ahmedabad, 50p.

NRSA, 1986. *Manual of Procedure for Wasteland Mapping Using Remote Sensing Techniques*. Govt. of India Publication, 52p.

NWDB, 1987. *Description and Classification of Wastelands*. Govt. of India Publication, 12p.

Pujahari, K.L., Sahu, D.K. and Mahapatra, G. 1980. Soil-scape of the lower Suklet irrigation project (Distt. Bolangir, Orissa) by remote sensing. *Proc. Semn. Application of Photo-Interpretation and Remote Sensing Techniques for Natural Resources Survey and Environmental Analysis, Dehradun*, pp. 158-160.

Rakshit, A.M. 1980. Geomorphic interpretations and mapping on Landsat imagery for diverse terrain types. *Proc. Semn.Application of Photo-Interpretation and Remote Sensing Techniques for Natural Resources Survey and Environmental Analysis, Dehradun*, pp. 110-112.

Ramesh, T.M., Maruthachalam, M. and Palanivelu, R. 1986. Digital temporal analysis of Landsat data for monitoring vegetation dynamics in a small watershed. *Proc. Int. Semn. Photogramm. and Remote Sensing for the Developing Countries, New Delhi*, Vol. 2, T- 5-P/17.1-17.7.

Rao, B.R.M. and Venkataratnam, L. 1985. Comparative study of Landsat imagery MKF-6 and Kate j 140 photographs obtained from Salyut space mission for soil resources mapping. *Proc. Six Asian Conference on Remote Sensing, Hyderabad*, pp. 544-551.

Reddy, M.B. and Gaikwad. R.L. 1986. Use of remote sensing techniques for targetting ground water in fractured crystalline rocks. Two case studies from Karnataka. *Proc. Six Asian Conference on Remote Sensing, Hyderabad*, pp. 322-732.

Sahai, Baldev *et al.* 1980. *Landuse Survey of Idukki District*. Technical Report, Kerala State Landuse Board, Trivandrum and Space Application Centre, Ahmedabad, 109 p.

Sangwan, B.S. and Garalapuri, V.N. 1980. Application of aerial photographic interpretation in the geomorphological and soil investigations of the south- western districts of Haryana State. *Proc. Semn. Application of Photo-Interpretation and Remote Sensing Techniques for Natural Resources Survey and Environmental Analysis, Dehradun*, pp.142-145.

Sanware, P.G., Karale, R.C. and King, G.J. 1985. Studies on landuse patterns and land degradation using landsat imagery. *Proc. Six Asian Conference on Remote Sensing, Hyderabad,* pp. 92-96.

Sehgal, J.L., Sharma, P.K. and Karale, R.L. 1985. Soil resource inventory of Punjab using remote sensing technique. *Proc. Six Asian Conference on Remote Sensing, Hyderabad*, pp. 298-301.

Sen, A.K. 1978. *Landuse classification System in Indian Arid Zone.* CAZRI Monograph, No. 9,44p.

Shankar, V. and Kumar, S. 1987. *Grazing Resources of Jaisalmer : Ecology and Development Planning with Special Reference to Siwana Grasslands*. CAZRI Monograph, No. 28, 92 p.

Shankarnarayan, K.A. and Singh, S. 1979. *Application of Landsat Data for Natural Resources Inventory and Monitoring of Desertification.* SDSU-RSI-VISP-79-17, U.SA. 136 p.

Shankarnarayan, K.A. and Singh, S. 1983. Application of Landsat data in determining the range biomass in Jodhpur district, western Rajasthan (India). *Journ. Ind. Society Remote Sensing* 10 (3) : 17-22.

Sharma, K.D., Singh, S.; Singh, N. and Bohra, D.N. 1989. Satellite remote sensing for mapping temporal changes in grazing lands. *Journ. Ind. Society Remote Sensing* 17 (4) : 55-59.

Sharma, K.D., Singh, S.; Singh, N. and Kalla, A.K. 1989. Role of satellite remote sensing for monitoring of surface water resources in an arid environment. *Hydrological Sciences Journal,* U.K. 34 (5) : 533-539.

Sharma, S.B. 1880. Application of small scale multiband imagery in coastal geomorphological studies. *Proc. Sem. Application of Photo- Interpretation and Remote Sensing Techniques for Natural Resources Survey and Environmental Analysis, Dehradun,* pp. 92-96.

Singh, N., Balak Ram; Kumar, S.; Kolarkar, A.S. and Saxena, S.K. 1989. Comparative study of TM and IRS-LISS II data for soil, landuse and vegetation resources in arid environmental. *Regional Workshop on Indian Remote Sensing Satellite Mission and its Application Potential, Jodhpur*, pp. 39-45.

Singh, S., Wasi Ullah, Ghose, B. and Kaith, D.S. 1969-71. Interpretation of aerial photographs of quantitative geomorphic characteristics of drainage basins in sub-humid to humid environments. *Ind. J. Geog.* 4, 5 & 6 (1) : 1-11.

Singh, S. and Kaith, D.S. 1971. Geomorphological analysis of landform features of Pali development block from aerial photographs. *Ind. J. Geog.* 49 (3 & 4) : 50-59.

Singh, S. 1976-77. A note on the photogeomorphic technique for locating ground water resources in Rajasthan desert. *Ind. J. Geog.* 11 & 12 : 83 - 84.

Singh, S. and Ghose, B. 1977, Geomorphology of Luni basin and its palaeoclimatic inferences. In : *Ecology and Archaeology of Western India* (Ed. D.P. Agrawal and B.M. Pandey). Concept Publ. Co., New Delhi; pp. 135-146.

Singh, S., Sharma, K.D., Singh, N. and Bohra, D.N. 1988. Temporal change detection in river courses and flood plains in an arid environment through satellite remote sensing. *Journ. Ind. Society of Remote Sensing* 16(4) : 53-56.

Singh, S., Ghose, B. and Kar, Amal. 1978. Geomorphic changes as evidence of palaeoclimate and desertification in Rajasthan desert, India (Luni development block - A case study), *Man and Environment* 2 : 1-13.

Singh, S. 1981. Geomorphology as an important tool in the exploration and planning of ground water resources in the Indian desert. *Development Policy and Administration View* 7(1) : 21-35.

Singh, S. and Shankarnarayan, K.A. 1982. Application of Landsat digital analysis in geomorphological investigations of the middle Luni basin, Rajasthan, India. *Ann. Arid Zone* 21 (4) : 279-287.

Singh, S. 1983. Application of remote sensing techniques in geomorphological investigations - A case study of Jodhpur district, India. *Ann. Arid Zone* 22 (4) : 283-294.

Singh, S. and Sharma, K.D. 1989. Application of aerial photo- interpretation of the morphohydrological characteristics of the drainage basins in sandstone region of Jodhpur district, western Rajasthan. *The Deccan Geographer* : 616-620.

Singh, S., Sharma, K.D., Singh, N. and Vats, P.C. 1989. Wastelands and rural development in the Rajasthan desert. *The Geographer* 36 (1) : 37-49.

Singh, S., Kar, Amal and Vats, P.C. 1991. Geomorphology of Rajasthan through remote sensing for environmental management. In : *Indian Geomorphology* (Ed. H.S. Sharma); Concept Publ. Co., New Delhi, pp. 269-292.

Singh, S. 1992. *Geomorphology and Remote Sensing in Environmental Management.* Scientific Publishers (India), Jodhpur. 281 p.

Singh, S. and Shankarnarayan, K.A. 1993. Remote sensing in monitoring temporal environmental changes in the Rajasthan desert. *Trans. Isdt. Ucds.* 8 (1) : 114-117.

Tomar, M.S. 1972. *Aerial Photo-Interpretation and Mapping in Pre-Investment Survey on Forest Resources.* Bulletin, Preinvestment Survey of Forest Resources Organisation, Dehradun, 110p.

Venkateswarlu, J. *et al.*, 1990. *Water 2000 AD - The Scenario for Arid Rajasthan.* CAZRI, Jodhpur.

Whitmore, G.,D. 1952. The development of photogrammetry. In : *Manual of Photogrammetry* (Ed. American Society of Photogrammetry), George Banta Publishing Co. Menasha, Wisconsin, pp. 1-16.

White, M.E. and Morain, S.A. 1979. *Evaluation of Landsat Based Area Measurement Accuracies for Surface Water Area in New Maxico.* Technical Completion Report No. 3109-216, New Mexico, Water Resources Research Institute, Las Cruces, New Mexico, U.S.A.41 p.

Yadav, R.C., Murthy, K.N.K. and Abichandani, C.T. 1976. Runoff potential of desert catchments in Bikaner district. *Ann. Arid Zone* 15 (1 & 2) : 1-7.